Remediation Plans and Templates

Published by CBRNE Ltd

i

Title:	CBRN Remediation Plans and Templates	
Date:	August 29, 2014	
Author(s):	N Hale	CBRNE Ltd
	D Kelly	CBRNE Ltd

This project has received funding from the European Community's Seventh Framework Programme. The views expressed in this document are purely those of the writer and may not in any circumstances be regarded as stating an official position of the European Community.

Front Cover Design by: Carolyn Smith BA (Hons) Ind Des MFA - CBRNE Ltd Design Director

Contents

2

1. Executive Summary

Following a terrorist incident involving CBR materials or an accidental release of similar hazardous materials there is likely to be a requirement to undertake some form of remediation[1] and/or disposal of assets[2].

This document is Deliverable D5.12 of Project PRACTICE. It introduces a set of plan templates and guidance notes which organisations may use to help them to plan for and obtain acceptance for such remediation tasks.

Three plan templates are provided, namely;

- a Remediation Plan template – used to set out the overall plans for the remediation and to gain initial acceptance of those plans from public and official bodies;

- a Remediation Justification template – used to demonstrate and gain approval for the detailed remediation plans, including the selection of remediation techniques; and

- a Remediation Confirmation template – used to gain acceptance for the completed works and authorisation/agreement to release the affected site from any further controls.

The plans will be particularly useful where the incident or the remediation has some potential impact upon the public. In these cases there may be oversight or supervision of the removal of the contamination hazard by a government or other official body.

It is intended that the plans should be used by organisations who have not previously had any experience with specialist remediation/decontamination and who may not have processes and arrangements in place for the management of such activities. The plans are designed for use in instances where, for example, a Local Authority or other controlling body has some responsibility for ensuring that the remediation is appropriately planned and controlled, but the owner of the assets is responsible for the actual remediation. In instances where the responsibility for remediation is assumed by government, or other national authorities, the plans may not be entirely appropriate but they may still provide instructive information to the organisation that has suffered the contamination. Similarly where an organisation is not responsible to a controlling body with powers granted by statute but is responsible to other Stakeholders the plans will still prove useful.

The plans may not be suitable where there are larger socio-economic or environmental issues that need to be addressed or where Government / State bodies undertake the works

---

[1] Within this Deliverable, Remediation is part of the Recovery Phase. It refers to the treatment of contaminated assets such that they can be safely re-used or disposed of; it therefore relates principally to decontamination.

[2] For example, buildings, office equipment, machinery, vehicles and IT equipment. For the purpose of this Deliverable, people and animals are excluded from the definition of Assets. Land remediation is included in principle but it is likely to attract much more regulatory control – as land contamination issues are widely covered by EU Directives – and additional assessments such as Environmental Impact Assessments will be required.

(rather than the organisation). Even in these cases however, it is suggested that the plan templates will represent a useful starting point and their acceptance, prior to an incident, by the parties involved will help to expedite the works.

It is proposed that an organisation who wishes to pre-plan for dealing with a potential incident of the type covered here should discuss the layout and content of the templates with its stakeholders and obtain consent for using them in the event of a real incident. Guidance on Stakeholder issues relating to contamination can be found in PRACTICE Deliverable D5.6 (Hale et al). Such approval and agreement before an incident will help to ensure timely completion of the documents and the associated remediation activities and will thus help to minimise the impact of the incident.

The three stage approach presented here is consistent with that developed and used in other similar instances, such as the decommissioning of radioactively contaminated buildings and facilities, the remediation of sites that have been contaminated during the illegal manufacture of drugs (such as methamphetamine) and the remediation of mould contamination in flooded buildings.

2. Glossary

The following terms are used with meanings provided below

Asset	A physical object belonging to or of value to an Organisation.
CBR	Chemical, Biological, Radiological
CWA	Chemical Warfare Agent
Dose	A measure of the degree to which an individual or object has been affected by a CBR agent (e.g. the amount of radiation received, the amount of chemical ingested etc)
Remediation	The treatment of assets etc such that they can be handled safely.
Stakeholder	A person, body or organisation that has an effect upon, or is affected by, the operations of an Organisation[3].
The Authority	Refers to the entity or group of entities that have some formal mandate (from government or law) to control the remediation works. This will usually be because public health may be at risk.
The Organisation	Refers to the party whose assets have been affected by the incident and whom is normally responsible for their remediation

---

[3] See Hale et al 2012 for further guidance on involvement of Stakeholders and their identification

3.   Introduction

This document is Deliverable D5.12 of Project PRACTICE. It introduces a set of plan templates and guidance notes which organisations may use to help them to plan for the remediation of assets following an incident or an attack involving CBR materials.

Three plan templates are provided, namely;

- a Remediation Plan template – used to set out the overall plans for the remediation and to gain initial acceptance of those plans from public and official bodies;

- a Remediation Justification template – used to demonstrate and gain approval for the detailed remediation plans, including the selection of remediation techniques; and

- a Remediation Confirmation template – used to gain acceptance for the completed works and authorisation/agreement to release the affected site from any further controls.

It is intended that the templates should be used by organisations who have not previously had any experience with specialist decontamination and who may not have processes and arrangements in place for the management of such activities. The plans are designed for use in instances where, for example, a Local Authority or other controlling body has some responsibility for ensuring that the remediation is appropriately planned and controlled, but the owner of the assets is responsible for the actual remediation. In instances where the responsibility for remediation is assumed by government, or other national authorities, the plans may not be entirely appropriate but they may still provide instructive information to the organisation that has suffered the contamination. Similarly where an organisation is not responsible to a controlling body with powers granted by statute but is responsible to other Stakeholders the plans will still prove useful.

The plans may not be suitable where there are larger socio-economic or environmental issues that need to be addressed or where Government / State bodies undertake the works (rather than the organisation). Even in these cases however, it is suggested that the plan templates will represent a useful starting point and their acceptance, prior to an incident, by the parties involved will help to expedite the works.

The need for documents such as these - and the stepwise approach which is presented - has been demonstrated through industry experience in gaining approvals for similar tasks such as the decommissioning of radioactively contaminated facilities, clean-up of illegal drug laboratories, mold removal in buildings and from ex post reviews of previous incidents (see References in Section 6). In particular they address the need for organisations to obtain stakeholder agreement to proposed actions and success criteria before work is started.

It is intended that an Organisation wishing to reduce the impact of such an incident on their activities should, **before the incident occurs**, obtain Stakeholder[4] agreement to the proposed content of and use of these templates if an incident does occur.

---

[4] The identification of Stakeholders is discussed in PRACTICE Deliverable D5.6 – see Hale et al.

7

In all cases, timely preparation of the plans using the templates will be aided by close co-ordination and communication between the First Responders, the Organisation, the Authority and other Stakeholders from the outset. Thus it is implicit in obtaining agreement to the format and content of the plans with the Stakeholders, that they also agree to provide the required support. This will be especially important with respect to the provision of data from the First Responders and others engaged in early stages of the incident. Organisations may wish to capture such agreements in a "Heads of Agreement" or "Memorandum of Understanding" as a separate document.

The following section provides a brief overview of Remediation as an introduction to the plan templates and guidance notes which are presented in Parts A to D of this document. It is intended that organisations wishing to use the templates will adapt them to their own formats and management systems. In particular they could form the basis of continuity management plans for such incidents or be integrated into existing incident management arrangements.

4.   Overview of Remediation

Following an accidental release of hazardous materials in the vicinity of an organisation, or a deliberate attack using such materials directed at an organisation, the initial Response phase involves removing the immediate hazard to health and protecting people from further immediate harm.   This is the phase where the emergency services[5] are active. It is likely that following the Response phase there will remain some contamination of assets such that they are no longer available for normal use[6].   The Recovery[7] phase is where the organisation goes about the tasks necessary to move on from Response and back to some form of normality[8]. Part of that Recovery may involve dealing with the contaminated assets such that they become safe.   This part of the Recovery phase is referred to here as Remediation (see Figure 1).

---

[5] In this context these may be the national/state Fire, Police or Ambulances services for example or they may be emergency services provided by a site or facility owner themselves – e.g. an on-site fire / rescue service.

[6] As the contamination may still present a hazard if the equipment were to be used normally, even if it presents little hazard in its quiescent state.

[7] Recovery is a broad term that encompasses all of the activities necessary to reach the new end state, such as retraining, restocking, relocation, recovery of contaminated assets, decontamination etc.

[8] This is not to say that it is not appropriate for an organisation to aim for complete recovery, but that they should recognise that it may be impracticable or even inappropriate. Indeed, some organisations may actually see an incident of the type considered here to be the opportunity or motivation to do something different or to change the structure or nature of the business; in the event of a significant incident that affects many organisations in an area, the business model in that area after the incident might well need to be different anyway; the customer base may change, the demand for certain types of goods may change etc. (Alesch)

Thus we are led to an overall picture of Recovery as moving towards some new end state and viewing Recovery as successful when that end-state has been achieved as intended. The definition of the end-state is therefore crucial to the problem of defining success or failure.

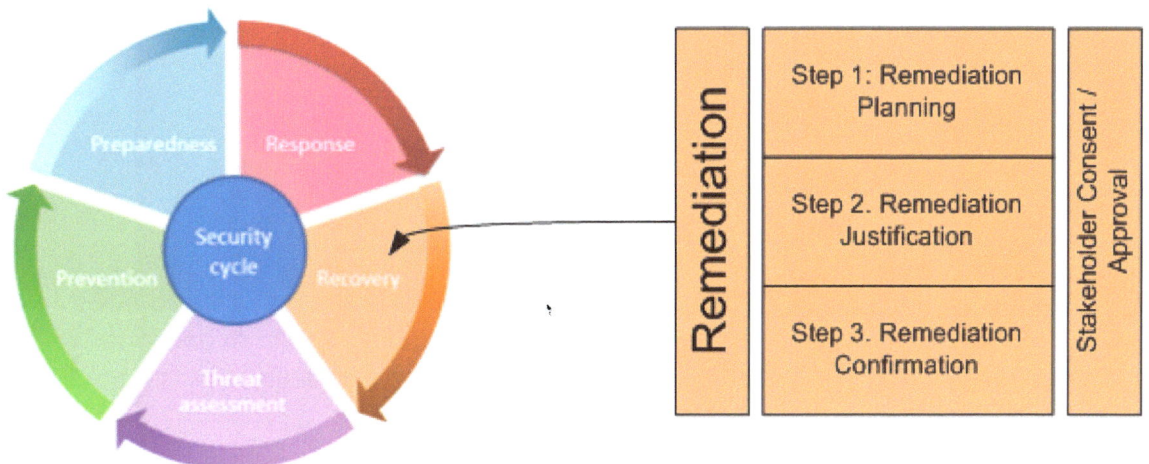

Figure 1: Remediation as Part of Recovery

In the absence of any external imperative, an affected organisation has the following choices[9];

> ➤ To isolate contaminated assets from further use or harm[10]
> ➤ To remediate them to permit their safe disposal (as waste)
> ➤ To remediate them to permit their re-use
> ➤ To remediate them to permit their sale

The choices made, determine the remediation goals. Thus, Remediation is goal oriented and may mean differing things to differing organisations depending upon their overall objectives for Recovery. Furthermore, it may mean differing things with respect to differing assets depending upon the intended destiny of those assets and/or their physical nature (which affects the degree to which they may be remediated). The level to which remediation is acceptable may also differ among and between stakeholders, organisations, and authorities.

These observations are fundamental to the templates that have been produced as part of this deliverable.

5. Overview of the Templates

The overall strategy presented in the templates is one of preparing progressive (or staged) documents that seek progressive permissions to proceed with stages of the project. This is consistent with that used widely within hazardous industries such as nuclear and petrochemical. It is also consistent with recommended best practice for dealing with situations like remediation of illegal drug laboratories and the like (see example references in Section 6).

The templates that are provided are for:

---

[9] The 'do nothing' option is, for the purposes of this document, ignored.
[10] This can include partial remediation to reduce but not totally remove the hazard.

- A Remediation Plan (Step 1 of Figure 1)

- A Remediation Justification (Step 2)

- A Remediation Confirmation (Step 3)

The templates are presented in Parts A , C and D. Part B of this deliverable contains an example of the use of the Remediation Plan template, for a theoretical incident.

For complex remediation tasks, each of the three plans should be prepared separately, in sequence. For more straightforward remediation tasks, where the approvals being sought are similarly simple, it may be possible to combine Steps 1 and 2 into a single document or to submit them at the same time without risking stakeholder agreement or technical accuracy[11]. Step 3 will always need to be a separate document.

The following sections provide an overview of each of the proposed plans and also provide guidance on when they should be prepared.

5.1 Remediation Plan [Step 1 of 3]

5.1.1 Purpose

The Remediation Plan sets down the specific details of the incident which has occurred. It provides sufficient detail to the responsible authorities for them to agree that the organisation may proceed with detailed planning for the remediation activities. Its key purpose is to set down the scope of the remediation activities to be undertaken, the control principles to be used by the organisation (to control management and health and safety risk, for example) and the criteria to be used to assess completion and safety. It sets out the legislation applicable to the hazardous materials introduced by the incident and relating to any resulting waste materials.

The Remediation Plan does not specify detailed arrangements for the actual execution of the remediation tasks as this is covered in the Remediation Justification discussed below.

5.1.2 Timing

The Remediation Plan presents the initial proposals for specific remediation following an incident. Therefore, it should be prepared after an incident has occurred, although parts of it can in principle be drafted prior to an incident as they are more generic in nature. The task of adjusting the Plan to the specific incident can commence as soon as information becomes available and thus its preparation can commence during the Recovery phase.

---

[11] For example, if the size of the work package is small, or similar situations have been dealt with previously without adverse stakeholder or official comment, or the science and technology being used are well established and accepted across the stakeholder group, it may be possible to include both the proposed criteria and methodologies into a single document.

For certain high risk establishments, like chemical sites, nuclear power plants and biological laboratories, general arrangements for decontamination may already exist in contingency plans and the like. These should be included in the template by the Organisation.

5.1.3 Data Sources

Key data relevant to the preparation of the Remediation Plan will be those arising from the incident and the actions of those involved in the immediate response. In many cases the Organisation will not have been directly involved with dealing with the incident nor will they have direct access to these data. As such it will fall to an official body (the Authority) to take responsibility for the co-ordination and provision of these data from the relevant parties to the Organisation. In particular, data regarding the potential scope of contamination will not be available to the Organisation at this time. Similarly, it is unlikely that the Organisation will have access to relevant health and safety data for some of the contaminants that may be involved with these types of incidents[12]. Again it will be relevant for the Organisation to seek guidance from the Authority or specialist advisers.

5.2   Remediation Justification [Step 2 of 3]

5.2.1 Purpose

The Remediation Justification sets out exactly what is to be done and how it is to be executed and controlled. It therefore provides the detailed substantiation of the acceptability of the proposed works by reference to items such as the criteria to be used (e.g. remediation standards, legal framework), the actual remediation techniques to be applied and how they will be managed and controlled. The aim of the Remediation Justification is to present to the controlling responsible authority sufficient evidence such that they are assured that the works to be undertaken will present an acceptable risk to all those concerned and that the works will comply with all relevant legislation.

5.2.2 Timing

The Remediation Justification presents the detailed arguments and plans for the remediation works. It is therefore prepared after an incident has occurred, when the actual hazards are known and the scope of remediation that is required is understood. Its preparation may commence concurrent with the Remediation Plan but its approval and authorisation can not be completed until after the Remediation Plan has been agreed with and approved by the responsible authorities.

In the event that the Organisation has already established arrangements for remediation with a third party – e.g. in the form of a call-off contract or framework agreement – then the details could be included in the template prior to an incident[13].

5.2.3 Data Sources

By the time that the Organisation prepares the Remediation Justification it will already have obtained much of the relevant background data and presented it in the Remediation Plan.

---

[12] For some contaminants these data may not yet be embodied in any formal regulations or standards due to their uncommon occurrence.

[13] Such call-off contracts and arrangements can significantly streamline the process of remediation.

Most of the information presented in the Remediation Justification will be generated directly by the Organisation or its subcontractors / specialist advisers. Where the Organisation intends to sub-contract some or all of the remediation works the template can be used by the Organisation as part of its specification for the scope of information to be supplied by them. Depending on the status of information supplied by the Authority at the Remediation Plan stage there is still the potential for further data to be supplied by them at this stage.

5.3 Remediation Confirmation [Step 3 of 3]

5.3.1 Purpose

The Remediation Confirmation provides a summary of the remediation works that were undertaken under the aegis of the Remediation Justification. Its primary purpose is to demonstrate that the works were completed in accordance with the Remediation Justification and to present the evidence that the residual risk from the incident has been reduced to acceptable levels. It also presents any lessons that may have been learned during the works such that these may be incorporated into the planning for future incidents.

5.3.2 Timing

The Remediation Confirmation is presented after the remediation works are complete and the required data are available. Its preparation may be commenced once the Remediation Justification has been prepared but it may not be completed until the works are complete and all wastes have been disposed of.

5.3.3 Data Sources

By the time that the Organisation prepares the Remediation Confirmation it will have completed the remediation works and between itself and its subcontractors will have generated much of the information required for this document.

In the event that the Authority procures its own independent evidence of satisfactory remediation, then it will clearly be responsible for the provision of those data to the Organisation who may choose to include them in the Remediation Confirmation or to refer to them.

Again, the template can be used by the Organisation to help produce a specification for its sub-contractors.

6. Literature

(ABS) Academy of Building Science (2008), *Mold Remediation - Qualifications & Proposal Evaluation*, available on line at http://www.atlantamold.biz/images/15%20-%20Proposal%20Evaluation%20Guidelines.pdf.

Alesch D.J Holly L.N, Mittler E and Nagy R (2001), *Organizations at Risk: What Happens When Small Businesses and Not-for-Profits Encounter Natural Disasters*, Public Entity Risk Institute, 11350 Random Hills Road, Suite 210 Fairfax, VA 22030

CAM Environmental Services (2005), Mold Remediation Protocol: A Guide for Contractors, Available at www.cam-enviro.com.

City of Westminster (2007), *Framework strategy for dealing with radioactive contamination arising from the circumstances surrounding the death of Alexander Litvinenko*, Westminster City Council.

Clarke, L (1989), *Acceptable Risk? Making Decisions in a Toxic Environment*, University of California Press ISBN 0-520-07657-5

(EPA) US Environmental Protection Agency (2009), *Voluntary Guidelines for Methamphetamine Laboratory Cleanup*, EPA-530-R-08-008, United States Environmental Protection Agency, Office of Solid Waste and Emergency Response. available at http://www.epa.gov/oem/meth_lab_guidelines.pdf.

Hale N, Kelly D (2012), Protocols for the Justification of Risk from Residual Contamination, Project PRACTICE, available at http://practice.fp7security.eu/downloads.

 (IAEA) International Atomic Energy Agency, (1999), *Decommissioning of nuclear power plants and research reactors: safety guide No WS-G-2.1*, International Atomic Energy Agency, Wagramer Strasse 5, P.O. Box 100, A-1400 Vienna, Austria.

Indiana Department of Environmental Management (undated), *Remediation Completion Report Completeness Checklist - State Form 54168 (1-10)*, Indiana Department Of Environmental Management Office Of Land Quality Voluntary Remediation Section, 100 N. Senate Avenue, MC 66-30, IGCN 1101, Indianapolis, In 46204-2251

Indiana Department of Environmental Management (2010), *Risk Integrated System of Closure (RISC) User Guide (June 2010)*, available at http://www.in.gov/idem/files/riscuserguide.pdf accessed September 2012.

National Academy of Sciences (NAS) (2005), *Re-opening Public Facilities after a Biological Attack – a Decision Making Framework*, National Academies Press.

(NEB) National Energy Board Canada, *Remediation Process Guide*, Available at www.neb-one.gc.ca/clf-nsi/rsftyndthnvrnmnt/nvrnmnt/rmdtnprcssgd/rmdtnprcssgd-eng.html#s13_1, viewed September 2012.

(NSD) Nuclear Safety Directorate (2005), *Technical Assessment Guide Guidance on the Purpose, Scope and Content of Nuclear Safety Cases - Issue 1,* available at http://www.hse.gov.uk/nuclear/operational/tech_asst_guides/tast051.pdf, accessed September 2012.

REMEDIATION PLAN FOR [*LOCATION*]

PREPARED BY [*ORGANISATION*] STEP 1 OF 3

APPROVED BY [*AUTHORITY*]

REVISION [??]

→ STEP 1: REMEDIATION PLAN

STEP 2: REMEDIATION JUSTIFICATION

STEP 3: REMEDIATION CONFIRMATION

RECORD OF APPROVALS

ROLE	SIGNATURE	PRINT	DATE
AUTHOR			
APPROVED (ORGANISATION)			
APPROVED (AUTHORITY)			

Contents

Guidance to Authors for the Completion of this Remediation Plan

1. The following text strings, which appear in italicised text in this document, must be replaced with appropriate text throughout. Some of these replacements may be made prior to an incident but others will only be relevant once an incident has occurred.

Text String	Description
[Date]	The date of the incident which caused contamination of assets that now require remediation.
[Organisation]	The name of the organisation (company, entity etc) with responsibility for the assets and for ensuring their remediation to the satisfaction of the Authority.
[Location]	The Location of the Site(s) at which the contaminated assets are located
[Hazardous Material]	The name of the hazardous material which was principally involved in the incident and which has caused contamination of assets that now require remediation.
[Authority]	The name of the Authority with legal responsibility for ensuring that the remediation works are satisfactorily completed (e.g. a Local Authority, District Authority, State Office, Government Agency etc)

2. Text that appears at the start of a section like this example is overview guidance and should be deleted once the required text has been inserted.

3. Text that appears in normal black font is more detailed guidance as to the topics that should be addressed to satisfy the requirements of the section. This text should be deleted once the template has been populated.

4. Text that appears {like this example} is example text that may be kept or modified as necessary by the Remediation Plan authors.

5. The Section headings should not be modified although additions are permitted.

6. This page may be deleted once the plan is complete.

7. A flow chart is shown overleaf which shows the process of completing this plan.

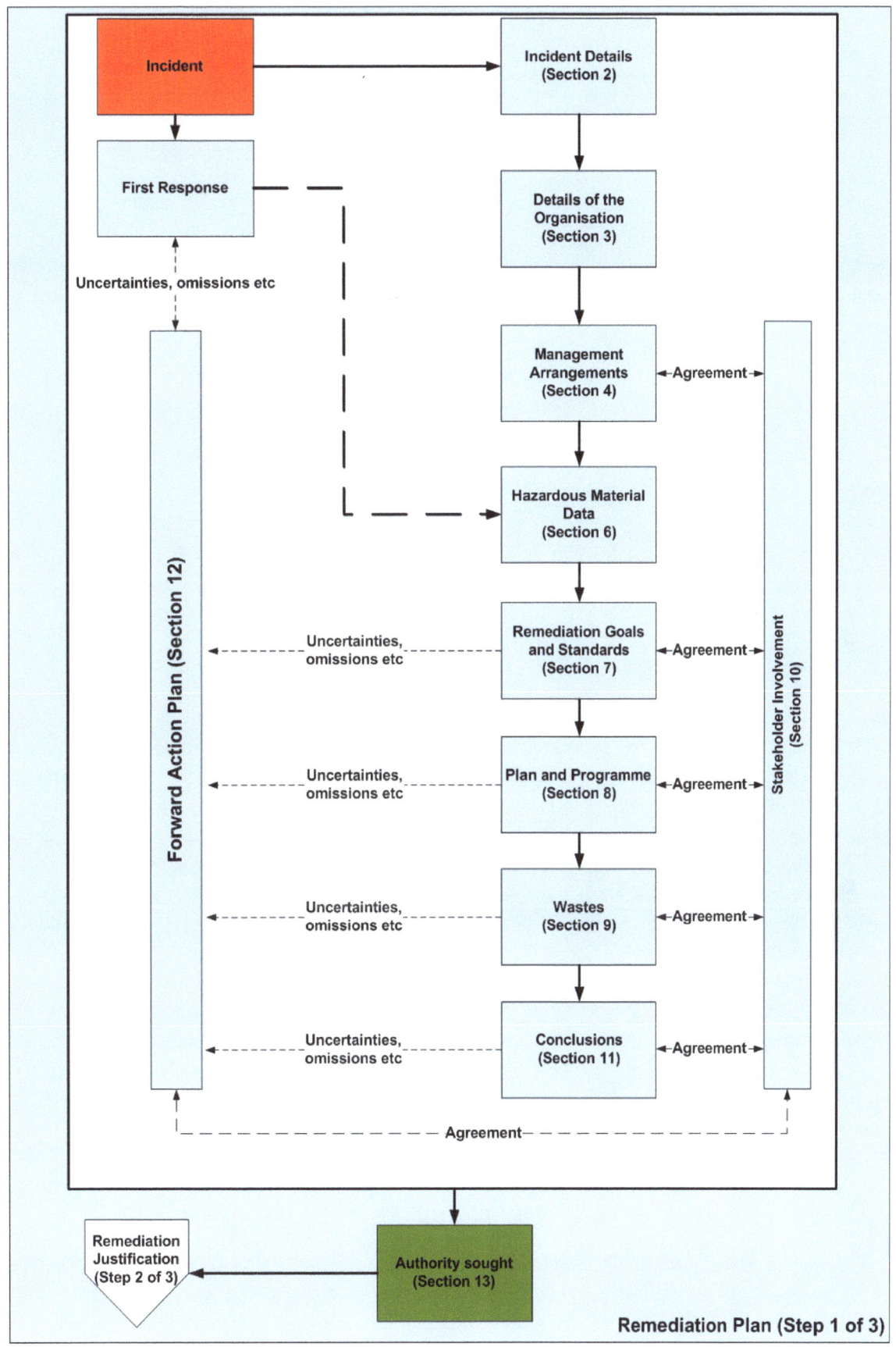

Flowchart for Completion of Remediation Plan

1.   Introduction

The key purpose of this document is to gain permission for the detailed planning of the required remediation.   Its early preparation will enable key issues to be identified with stakeholders and Authorities before any significant effort is expended on detailed planning. Parts of this template may be completed in advance with standard Organisation, Authority and Stakeholder details, for example.

The level of detail presented in the completed document should be commensurate with the complexity of the proposed tasks, their associated hazard potential and the hazard potential of the contaminants.

Many sections of this document will need input from third parties, such as those who were involved in the immediate response to the incident, or specialist bodies engaged to provide technical support and advice.

The main text should be kept as straightforward as possible with detailed technical information presented in the Annexes. In this way it is possible that the main document will be readable by and accessible to external third parties who may be interested but whom may not have detailed technical knowledge.

{On *[Date]*, some of the assets of *[Organisation]* at *[Location]* became contaminated as a result of an incident (the Incident) involving *[Hazardous Material]*.

This Remediation Plan is the first in a series of three stage submissions covering the required remediation works which will be produced. The purpose of this Remediation Plan is to describe the arrangements for remediation works required at *[Location]* that are to be undertaken by *[Organisation]* and to provide a vehicle by which *[Authority]* and *[Organisation]* may formally record their agreement to proceed with detailed planning for the works. Furthermore, it provides a single reference point for the relevant information that interested Stakeholders may refer to.

This Remediation Plan sets out the background to the incident which caused the contamination and outlines proposals for the control and execution of the remediation works. It does not define the actual methods of decontamination nor does it seek permission to undertake the works but just approval to progress to the detailed planning stage. Justification for the remediation works will be presented in a subsequent stage submission called a Remediation Justification. Similarly, justification for termination of the works and release of the site from any special controls will be sought through the final stage submission called a Remediation Confirmation.

Some additional sampling and analysis work might be undertaken prior to the work in this plan in order to gain sufficient data to enable it to be prepared. Nothing in this plan stops such works but they will be controlled under ad-hoc arrangements outside of the scope of this plan.

The following sections set out the background to the incident and its timescales, details of *[Organisation]* and the proposed arrangements for managing the works and the waste materials that may be generated by them. Furthermore it sets out a Forward Action Plan which will be used to manage and monitor key actions that remain to be undertaken.

This Remediation Plan has been prepared using data provided by the *[Authority]* and the *[Organisation]*.}

2. Details of the Organisation

> This section will provide legal details of the Organisation and its relevant activities
>
> This is important as it helps to define the importance of the assets to the organisation so that the priorities for remediation can be planned with this in mind. This information also helps the Authority and other Stakeholders to understand the significance of the remediation works to the Organisation and to place them in context. If possible, it may be helpful to note the financial implications of the loss of the site or loss of amenity caused by the incident.
>
> It is likely that much of this section can be prepared in advance of an incident.

State the legal name of the Organisation, its status and any registration numbers or other identifiers.

Provide a concise statement of the nature of the Organisation's business at the affected premises and present an organisation chart in Annex V.

Refer to any unusual legal controls which apply to the Organisation's business (i.e. if it is subject to special regulatory controls such as those in passenger transport).

3. Details of the Incident

> *This section will summarise the incident in terms of the hazardous materials involved, when it occurred, the overall extent of the resulting contamination etc. It will provide the reader with an appreciation of the overall scope and the history to date. (Technical details of the substances involved and the extent of the resulting contamination are presented later) Some descriptive text can be added to aid clarity if required.*
>
> Key data for this section of the Plan will come from the responding agencies and the authorities (e.g. emergency services and law enforcement agencies) as the Organisation will not normally have control over the early stages of an incident and will not normally have access to knowledge concerning CBRN contaminants. The Authority will be responsible for providing supporting information from, for example, police incident reports, emergency services, health care bodies etc.

Provide summaries of reports and data provided by external bodies (police, fire etc), copies of the reports should be compiled in Annexes this document.

If the incident has led to the contamination of sites for which other Organisations are responsible and for which separate remediation activities are being undertaken then reference to those activities should be made here.

Provide a detailed record of the incident's timeline in the Incident Timeline Summary Table below and summarise it here.

Explicitly name the hazardous materials introduced as a result of the incident.

Initial Location of Incident	Specify the point at which the incident started – this will be the location at which the hazardous material was initially released
Substances Involved	Name the substances involved
Weather at time of incident	Note the prevailing weather at the time of the incident – in case there are concerns regarding existing spread of contamination due to weathering.

Timeline		
Date	Occurrence	Comment
00/00/0000	Initial Event	Etc
00/00/0000	Emergency Services Arrival	Etc
00/00/0000	etc	Etc
00/00/0000	Etc	Etc
00/00/0000	Etc	Etc
00/00/0000	Police release of scene	Etc

Scope	
Areas affected	Name the areas affected that now need remediation
No. People directly affected	Identify the potential number of people who were directly affected by the incident – i.e. those in the immediate vicinity. Note numbers of casualties or who received hospital treatment [14]

3.1   Actions taken by Emergency Services

This section will describe the actions taken by the emergency Services following the incident.

Present a summary of:

i)   The names of the emergency Services that attended and what they did

ii)   The use of any chemicals or materials which may have changed or modified the nature of the contamination or its dispersion (e.g. any decontamination or neutralisation chemicals used by the Services).

---

[14] – Although this plan does not deal with decontamination of people, it is appropriate to present data explaining the impact of the incident on people at the time of the incident to ensure that the scope of the incident and its hazards are properly appreciated.

iii)   Any wastes or other materials generated by the initial responders or left by them at the scene[15]

iv)   If any of the site is now occupied or not

3.2   Actions taken by Other Agencies

This section will describe the actions taken by the other Agencies following the incident.

Describe any actions taken by Local Authorities, Law Enforcement and other agencies after the incident. Detailed reports should be compiled in Annex I.

Make particular note of any prohibitions, covenants or other restrictions placed on the Organisation or the Location as a result of the incident.

Record any materials removed or barriers erected.

Note: The Authority will normally collate and provide these data to the Organisation.

3.3   Extent of Contamination

This Section will provide information regarding the physical extent of the contamination – as known at the time of writing.   Known areas of uncertainty should also be highlighted.

Key data for this section will come from the responding authorities and agencies and will be derived from the safety zones and cordons set up during the crisis and from surveys and investigations during forensic examinations, for example. The data will require interpretation by the Organisation and will be supplemented by the Organisation with data pertaining specifically to the site (e.g. its size, shape, extent etc). Any data presented will be referenced back to its source.

At this stage it is unlikely that the detailed extent of contamination will be known, nor is it necessary that it is known at this stage.   The purpose of this section is to define the overall boundary or envelope of the contamination.

Insert drawings and plans showing the extent of contamination in Annex II and summarise it here. Refer to the mechanism by which the contamination was dissipated – it may be useful to refer to the Figure below as a means of considering all of the potential ways in which the contamination may have become a hazard to people.

---

[15] e.g. Waste protective clothing, forensic wastes etc.

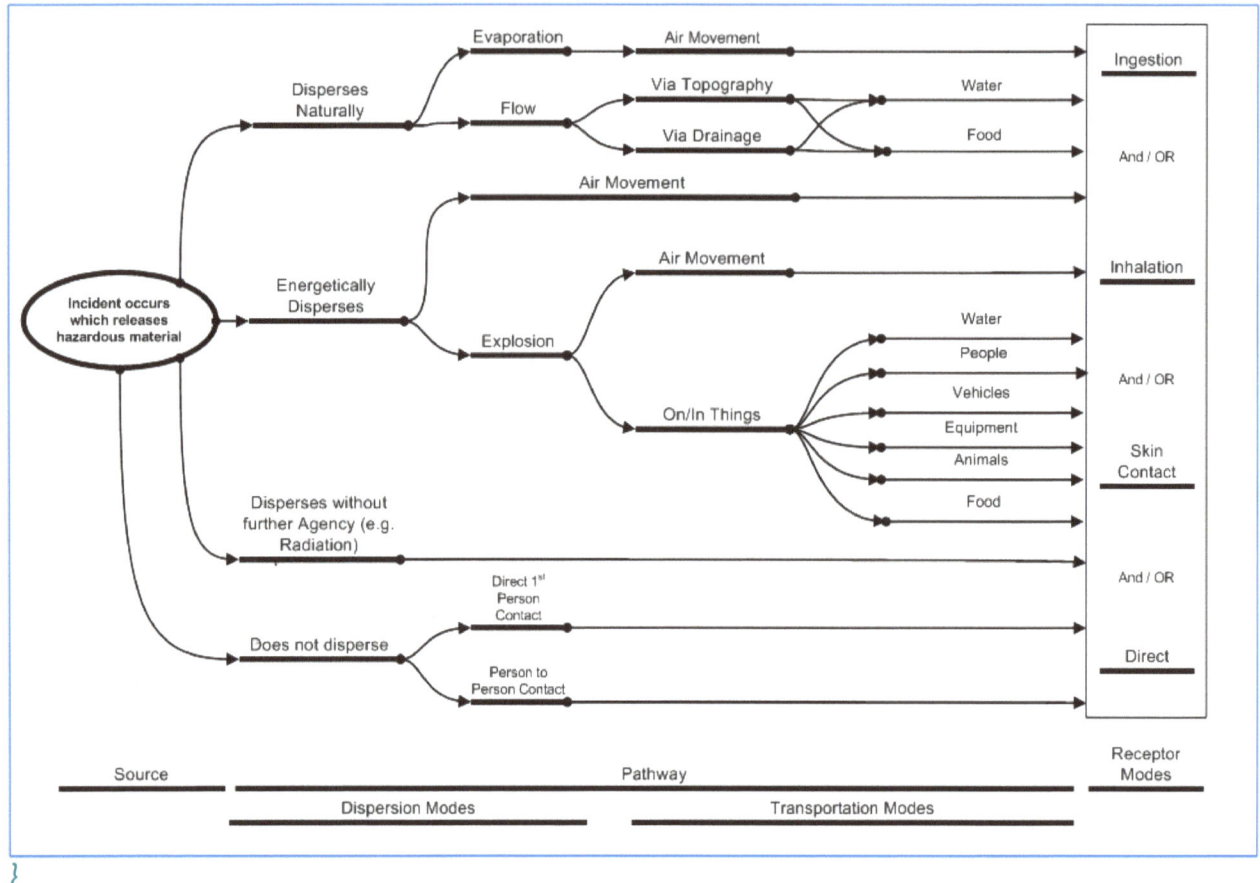

Present the best available data at the time of preparation of this Remediation Plan. If further sampling and investigation will be undertaken prior to and during the remediation works then state that here and add it as an action in the Forward Action Plan in Section 12.

4. Remediation Management Arrangements

This section will detail the proposed management arrangements for the remediation works. It is a key section of this Plan. Its key purpose is to show how the Organisation will ensure that it retains control of the remediation works and how the requirements of the Authority are to be met. It therefore provides confidence that the works will be executed in a controlled and safe manner.

Present a summary of the overall management arrangements that will apply to the remediation works.

The following key points should be noted;

i) A commitment that all works will be undertaken strictly in accordance with written method statements and risk assessments is likely to be required. The Authority may wish to approve these prior to them being implemented.

ii) Any third party arrangements (e.g. the appointment of a waste disposal contractor by a remediation contractor) will require explicit mention and control.

iii) Detailed management arrangements will be presented in the Remediation Justification document submitted later.

Present the names and contact details of the key contact points within each of the Organisation and the Authority in the Table below. Multiple contact points may be nominated - e.g. for separate key areas such as Quality Management, Commercial Matters and Technical Management - but a single point of contact is preferable.

Key Contact Points for the Remediation Works

Position	Name	e-mail	Tel No.
{Key positions	Name	email.address@a...	123456789}

Use the Table below to identify key roles and responsibilities for areas such as waste management, health and safety, environmental issues, commercial matters etc.

Key Roles for Remediation Works

Role	Organisation	Responsibility / Authority	e-mail	Tel No.
Description of Role (e.g. Health and Safety Manager)	[Organisation]	Define their key responsibilities and authorities for the remediation works	{email.address@a...	123456789}
Name of the key person in the Authority who is the point of contact}	[Authority]	Define their key responsibilities and authorities for the remediation works		
Names of other key contacts or co-ordinators, e.g Technical Advisers, Environmental Agencies, Specialist Contractors etc	etc	Define their key responsibilities and authorities for the remediation works		

4.1 Arrangements for Health, Safety, Quality and Environment

Insert a brief summary of the general arrangements for health and safety – reference to existing management systems and procedures should be used where these are relevant to the types of works covered by this Remediation Plan. Where the Organisation needs to make additional arrangements –such as the appointment of specialist advisers – then these should also be highlighted.

Note that detailed presentation of these arrangements should be deferred until the Remediation Justification Document is prepared but it is necessary to present sufficient detail here to show that the works can and will be appropriately controlled.

4.2   Arrangements for Control of Contractors

Insert a brief summary of the arrangements specifically relating to the control of Contractors – i.e. demonstrate how overall responsibility and control still rests with the Organisation, even though specialist contractors may be used for some of the actual works, As above, reference to existing systems is acceptable as long as these are shown to be relevant to the types of works covered by this Remediation Plan.

Detailed presentation of these arrangements should be deferred until the Remediation Justification Document is prepared but it is necessary to present sufficient detail here to show that the use of contractors can and will be appropriately controlled.

Note: The Authority may wish to formally approve the use of any sub-contractors prior to their engagement.

4.3   Health Management

Some contaminants require specialised monitoring in order to detect their impact on those exposed to them. This section is used to define those arrangements, where they are necessary. If no special arrangements are required then this should be explicitly stated.

Insert a brief summary of any arrangements that may be required for the monitoring and recording of the health of those potentially exposed to the contaminants (e.g. routine medicals, blood / urine testing, psychological monitoring). For the public and other external stakeholders (i.e. those not directly involved with the remediation works) these arrangements are likely to be advised by the Authority or their technical advisers. For remediation workers and other employees / contractors of the Organisation these arrangements will be the responsibility of the Organisation but will likely also be advised by the Authority and their technical advisers.

Detailed presentation of these arrangements should be deferred until the Remediation Justification Document is prepared but it is necessary to present sufficient detail here to show that the health of all those affected by the works can and will be appropriately controlled.

5.   Financial Arrangements

The purpose of this section is to show that there are sufficient financial resources available to complete the remediation works satisfactorily. It also helps to manage the expectation of stakeholders in terms of what is realistically possible. Mention should also be made of any advance arrangements on costs and resources made with contractors, e.g Framework agreements where levels and costs of contractor response are determined in advance of remediation works.

Define the financial resources available to the remediation project and where those resources are coming from. Define any limitations of those arrangements or any financial controls that may be introduced by insurance considerations.

Explicitly define where funds are being drawn from – e.g. internal reserves, disaster funds, insurance arrangements etc.

6. Hazardous Material Data

This section will present data regarding the nature and severity of the hazardous materials and the resulting contamination. It will also present data regarding any existing hazards on the site which may be of relevance to the remediation works – e.g. any existing contamination from previous industrial incidents or any hazards associated with the materials used at the site during the Organisations normal activities.

The degree to which the information can and should be presented in the Remediation Plan will depend upon the status of the remediation planning and will be a matter for the Organisation and the Authority to decide upon. It is not the purpose of this section to develop the methods by which these hazards will be addressed. These will be presented in the Remediation Justification Document.

6.1 Hazards Arising from the Incident

Provide a summary of the hazards resulting from the incident. It is recommended that these hazards are identified with - or are at least confirmed with - Stakeholder involvement – see PRACTICE Deliverable D5.6 'Protocols for the Justification of Risk from Residual Contamination' (Hale etc al) for further guidance and information. See also Section 10 of this document.

Focus should be on any hazardous material introduced by the incident - e.g. materials directly associated with the incident - but note should also be made of any materials which may have been introduced in response to the incident – e.g. any treatment chemicals, neutralising agents etc. Also make note of any changes in the hazards that may be introduced by the remediation works – e.g. those works may cause some of the contaminants to become more mobile.

It may be useful to refer to the Figure in Section 3.3.

6.2 Pre-existing Hazards

This section is used to present details of other significant hazards on the Site, not necessarily arising from the incident. This is required for completeness to ensure that they are not overlooked in the detailed planning.

Present data regarding any existing known hazardous materials that existed in the affected areas prior to the incident (e.g. asbestos, process chemicals, cleaning chemicals etc). These details are important as they may, for example, lead to additional considerations with regards to waste management or the use of the chemical decontaminating agents (which may interact with pre-existing contaminants).

7. Proposed Remediation Goals and Standards

The purpose of this section is to present the remediation goals and standards[16] that will be adopted i.e. the extent to which it is proposed to remediate the affected areas and why. At this stage of planning these will of necessity be high levels goals and standards but they will be developed in detail in the remediation Justification Document submitted later.

7.1 Remediation Goals

Notes:

i) There may be more than one goal/standard as the risk from different items may differ depending on their intended future use and location.

ii) It is not necessary to note the remediation methods here as they will be presented in the Remediation Justification Document; they will require detailed consideration and probably input from specialist sub-contractors.

iii) It is not necessary here to define which goals will be applied to which assets in detail as this will be done in the Remediation Justification Document. Only a high level discussion of the goals is required here.

At a minimum, this section should note the intended goals of the remediation - e.g. the Organisation wishes to:

➤ Recover some assets that are to be re-used or retained by the organisation, without controls (normal use) and/or

➤ Retain some assets under prescribed controls and safeguards (e.g. items may be kept in locked rooms or otherwise isolated from future use) and/or

➤ Dispose of some or all assets as hazardous wastes, and/or,

➤ Dispose of some or all assets (to controlled disposal sites) as non hazardous wastes, and/or

➤ Release Assets without any restrictions on their future use or destiny (so called free release).

It is important to be specific about the intended goals of the proposed remediation since the appropriate remediation standard may vary significantly with the goal. In addition, it sets the remediation plan in context with the commercial plans for the affected areas and for the business as a whole. Moreover, it makes it clear to all those involved what level of remediation is being sought – cross-refer to Sections 3 and 5 as appropriate.

---

[16] The remediation goals are the intended end-points (e.g. re-use, disposal etc). The standards are the evidence based measurements or observable facts that will be used to confirm that the goal has been achieved. Goals and standards therefore go together in pairs.

7.2 Remediation Standards

If appropriate remediation standards already exist for the contaminant and have already been specified by the Authority then they should be noted here in summary and presented more fully in Annex III. If an appropriate standard has not been identified then note that here and also that a standard will be presented and justified in the Remediation Justification Document – in this case add this action to the Forward Action Plan in Section 11.

The standards should be linked to the goals in the previous Section - i.e. there should be a standard for each goal.

8. Proposed Remediation Plan and Programme

This Section will present the outline programme for the works. This will be refined and confirmed in the Remediation Justification Document.

Present a bullet form overview of the key proposed phases of the remediation works. They key stages that lead to significant reductions in hazards or removal of key assets should be noted. The key hold points for the approval of the Remediation Justification and the Remediation Confirmation should be explicitly shown.

9. Wastes

This Section is used to outline the volumes and types of wastes that are anticipated for the works. These are key considerations in respect of waste management and environmental impact. A detailed waste management plan will be developed as part of the Remediation Justification document.

Itemise the expected waste arising from the remediation works. For each waste stream provide the following information or estimates.

- A waste stream description (e.g. solid, liquid, hazardous, non hazardous etc)

- A brief description of the stream (e.g. what it will be comprised of, what sorts of materials etc)

- An estimate of the quantity (kg, m^3, bags, skips etc)

- The waste route (i.e. where it is anticipated that the waste will be disposed of, e.g. landfill, waste incinerator, hazardous waste site etc).

- For hazardous wastes; the method of encapsulation or containment to be used (e.g. double polythene bags, steel drums, covered skips etc).

10. Stakeholder Involvement

This section will present the proposed arrangements for Stakeholder Involvement. The extent of these arrangements will be a matter for the Organisation and the Authority to agree and will depend largely upon the perceived significance of the incident to the community, the organisation's investors, the nature of the business etc. Good Stakeholder engagement at this early planning stage is likely to lead to better acceptance of the project and its results later (Hale etc al).

Provide details of how the Organisation intends to engage with Stakeholders for the project. Note: the Authority may wish to handle public Stakeholders.

Guidance is available in PRACTICE Deliverable D5.6 'Protocols for the Justification of Risk from Residual Contamination'.

11. Conclusions

{A remediation plan has been presented which sets out at high level the arrangements that are proposed by [Organisation] for the management and execution of the required remediation at [Site], in order to return the affected assets to the required state.

Detailed substantiation of the proposed remediation works will be presented in a Remediation Justification Document, to be presented after approval of this plan.

Arrangements for the remediation works will be developed in consultation with relevant Stakeholders.

Approval for release of the affected assets from controls will be sought through the production of a Remediation Confirmation Document which will also be submitted to [Authority] for approval.}

12. Forward Action Plan

The Forward Action Plan, presented here, is used to identify outstanding actions and items requiring further analysis or assessment in order to progress the project through to completion. The Forward Action Plan (FAP) will be progressed through the project and its status presented at each of the subsequent stage submissions.

After completion of the previous sections, collate below any actions necessary to ensure that any identified gaps or requirements are addressed.

Forward Action Plan

No	Action	Responsible Party	Timescale
1	Approve this Remediation Plan	*[Authority]*	
2	Prepare Remediation Justification Document	*[Organisation]*	
3	Approve Remediation Justification Document	*[Authority]*	
4	etc		

13. Authority Sought

This section clearly identifies the scope of permission sought from external parties.

Clearly set out the exact scope of the permissions being sought and from whom. Identify any hold-points or further permissions that may be sought –e.g. "Permission is sought from the Local Authority to commence detailed planning for remediation of the Site as set out above. Permission to commence remediation works on the site will be sought separately after completion of a Remediation Justification document."

14. Literature

I   Annex I: Reference Documents Describing the Incident

II   Annex II: Extent of Contamination

Insert drawings, survey reports etc showing the known extent and degree of contamination at the affected locations.. Where contamination of drains is suspected the extent and location of these drains and their intersections should be clearly shown.

{ e.g.

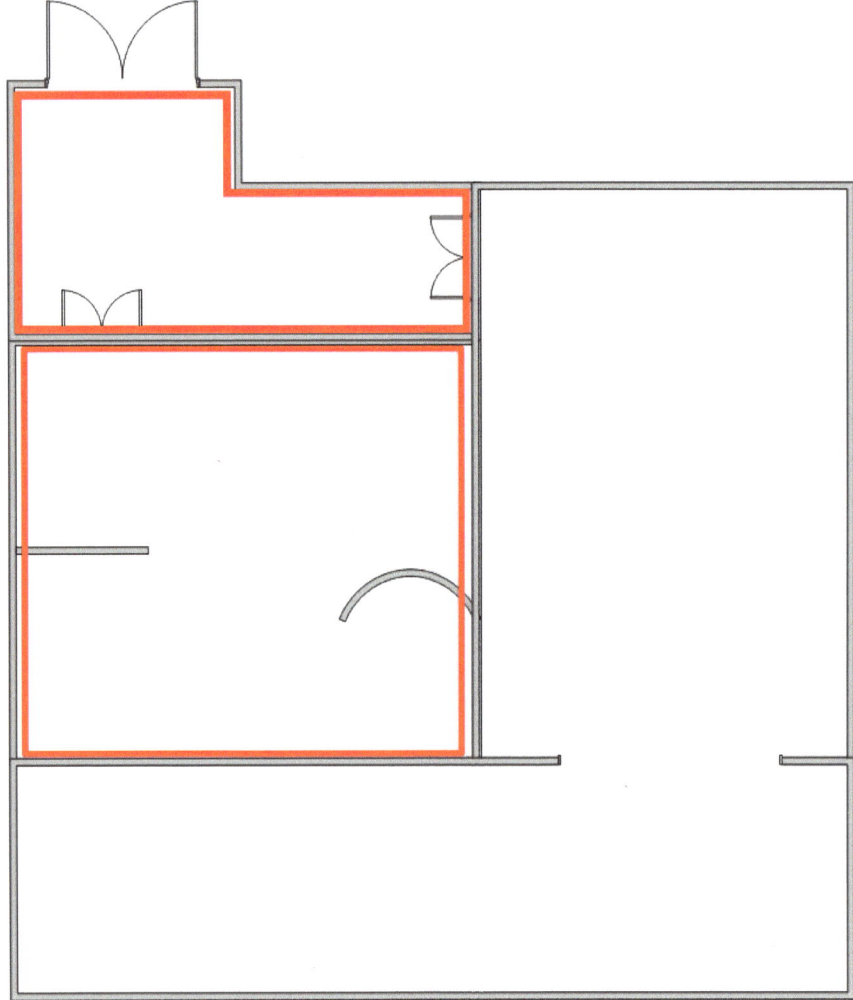

Figure AII.1: Extent of Contamination (In Red) }

III   Annex III: Details of Hazardous Materials

Provide a summary of the known data regarding the contaminants in the table below. Refer to or attach references for all data values. Complete the following table for each substance introduced by the incident. Use NK (Not Known) for any values which are not known at the time of writing.

Name(s) of the Contaminating Substance(s)		C	B	R
Substance name(s) – formal name and any other names by which it may be known.		☐	☐	☐
Existing Exposure Standards and Limits	**Method of Measurement**	**Value**		
Known Health Hazards				
Hazards				
Applicable Legislation				

IV   Annex IV: Legacy Hazardous Materials

Insert details of any pre-existing hazardous materials at the site which may have an impact on the remediation works

V   Annex V: Organisation Charts and Contact Details

Insert organisation diagrams showing the key lines of responsibility from the Board level to the operational unit affected by the incident. Show how the remediation itself will be managed; show the levels at which key decisions regarding the remediation works will be taken and the connections between the organisation and the authority.

e.g. {

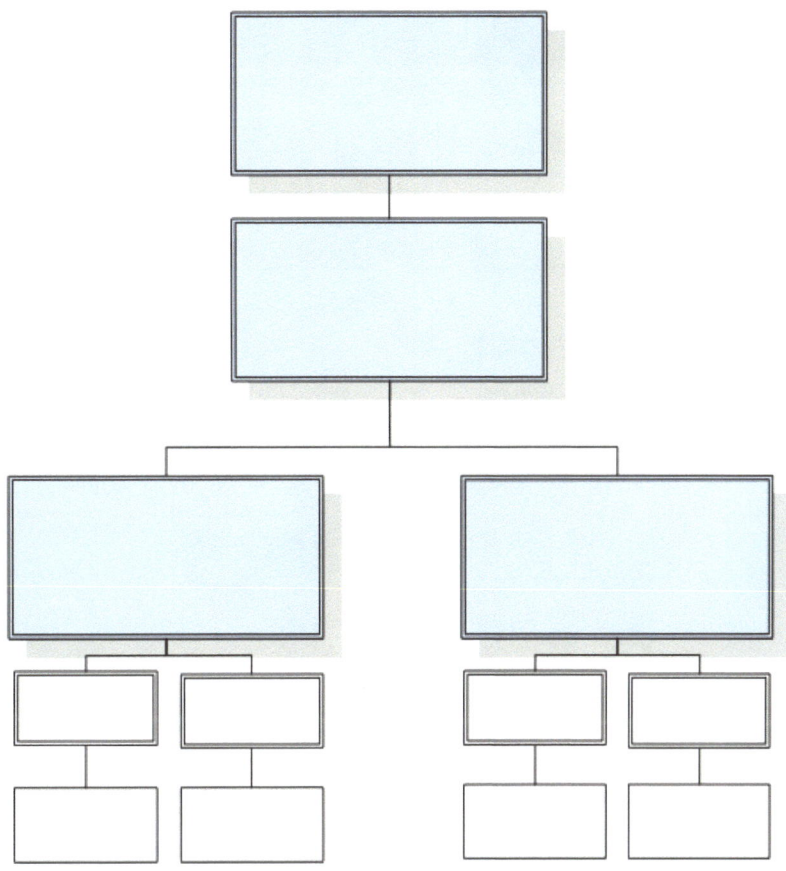

Figure A5.1 Organisation Chart}

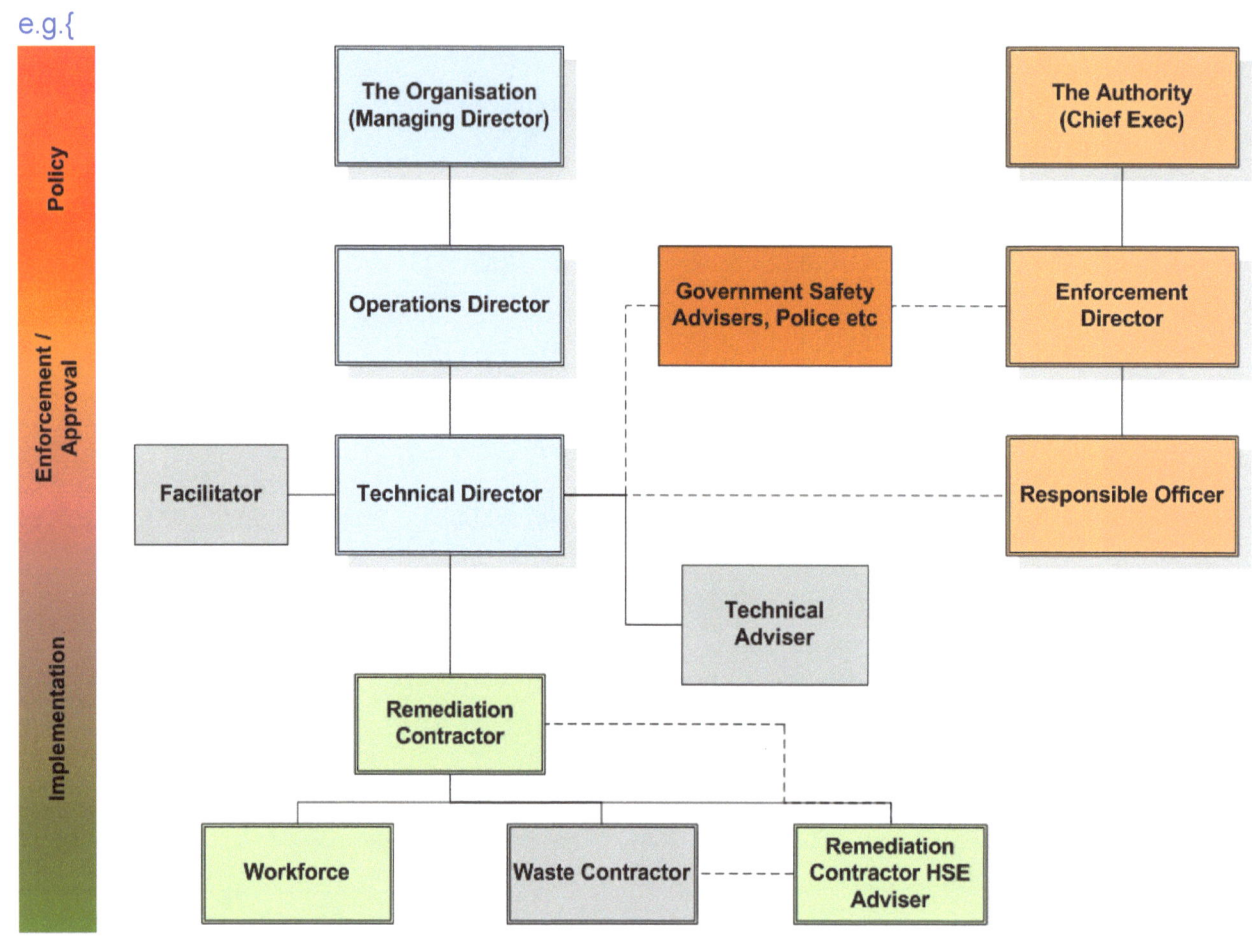

Figure A5.2: Remediation Management Organisation Chart}

Contact Details (from Section 4)

Position	Name	e-mail	Tel No.
{Key positions}	{Name}	{email.adrress@a...}	{123456789}
{Key positions}	{Name}	{email.adrress@a...}	{123456789}
{Etc}	{Etc}	{Etc}	{etc}

VI Annex VI: Details of Any Restrictions or Prohibitions

VII   Annex VII: Miscellaneous

REMEDIATION PLAN FOR CONCEN_HALL_1
PREPARED BY CONCENCO LTD
APPROVED BY NCC
REVISION [1]

RECORD OF APPROVALS

ROLE	SIGNATURE	PRINT	DATE
AUTHOR			
APPROVED (ORGANISATION)			
APPROVED (AUTYHORITY)			

Contents

1.   Introduction

(Note: This example Remediation Plan is based loosely upon the scenario developed for Project PRACTICE which is presented in D8.1.   Some details have been generalised or replaced to conform to the Project's security guidelines for unclassified documents.   The name ConCenCo is not intended to refer to any existing company, institution or entity; any resemblance to any such entity is entirely coincidental.   The whole of this document is fictitious and is only intended to provide further guidance to those wishing to complete a Remediation Plan).

On *1/4/2012*, some of the assets or ConCenCo Ltd in ConCen_Hall_1 (a conference hall) became contaminated as a result of an incident (the Incident) involving Niras[17]. ConCen_Hall_1 is a conference hall in a conference complex called ConCen.

This Remediation Plan has been produced in line with an Agreement between ConCenCo and NeuTown City Council (NCC) as set out in Reference 1.   Its purpose is to describe the arrangements for remediation works required at ConCen_Hall_1 that are to be undertaken by ConCenCo and to provide a vehicle by which NCC and ConCenCo may formally record their agreement to proceed with detailed planning for the works.   Furthermore, it provides a single reference point for the relevant information that interested Stakeholders may refer to.

This plan sets out the background to the incident which caused the contamination and outlines proposals for the control and execution of the remediation works. It does not define the actual methods of decontamination nor does it seek permission to undertake the works but just approval to progress to the detailed planning and execution stages. Justification for the remediation works will be presented in a subsequent Remediation Justification Document. Similarly, justification for termination of the works and release of the site from any special controls will be sought through a Remediation Confirmation Document.

Note that some additional sampling and analysis work might be undertaken prior to the work in this plan in order to gain sufficient data to enable it to be prepared. Nothing in this plan stops such works but they will be controlled under ad-hoc arrangements outside of the scope of this plan.

---

[17] A fictitious substance

2.   Details of the Organisation

ConCenCo Ltd (company registration number 1111999) is an entertainment facilities company specialising in providing training facilities and small conference halls to the public and industry. The ConCen site is one of five owned and operated by ConCenCo Group (head office in Paris) who wholly own ConCenCo Ltd.

ConCenCo Ltd employs approximately 150 people at the ConCen site. ConCenCo Ltd is one of the largest employers in the immediate area.

The Managing Director of ConCenCo Ltd reports directly to the European Director of ConCenCo Group in Paris (see Figure A5.1).

The building and its contents are wholly owned by ConCenCo Ltd.

Hall 1 is responsible for approximately 25% of the turnover of ConCenCo, which was €10million in 2006.

3.   Details of the Incident

The following sections provide an overview of the incident which occurred at the ConCen site. Data have been drawn from information provided by NCC and ConCenCo's experience of the incident. Detailed references are presented in Annexes.

A detailed record of the incident is presented in Table 1 and is summarised below.

On 1/4/2012 at 13:00 CET a religious group released approximately 2 litres of Niras mixed with an unknown solvent in ConCen_Hall_1 (one of the conference halls) at the ConCen site, during a religious conference. The Niras was released using a simple mechanical mechanism (the details of the actual method of release are confidential). This resulted in both liquid splashes and the release of vapour and spray.

Several members of the audience in the hall suffered respiratory distress shortly after the incident. The emergency services were called and the building was promptly evacuated. The perpetrators were not arrested as they were not present during the incident. One elderly person died from exposure to the chemical and five other persons remain in hospital receiving treatment.

The weather was dry and cool with a weak westerly wind.

3.1   Actions taken by Emergency Services
The incident was initially attended by fire, ambulance and HazMat response teams who ensured that the building was evacuated, cordoned off and that the affected people were decontaminated with soapy water and removed to the local hospital for treatment.

The building ventilation systems were isolated by the emergency services to restrict further spread of vapour (under the guidance of the ConCen maintenance staff).

Although the decontamination run-of was largely captured by the decontamination units, some contaminated run-off into the site drains is suspected to have occurred.

As a precautionary measure the whole ConCen site was evacuated – although certain parts of the conference centre are now in use again.

It is understood that the emergency services applied 'Fullers Earth' to some obvious areas of liquid contamination and also that some alkali neutralising agents were used (BX90). Some contaminated items were bagged by the services and have been left in the Hall. The exact contents of these bags of waste is unknown (see Section 9 "Wastes").

3.2 Actions taken by Other Agencies

Because of the deliberate nature of the incident and the fatalities, the police were also involved. Since the attack they have entered ConCen_Hall_1 on several occasions to remove forensic evidence. The mechanical device has been removed by the police.

Due to the hazardous nature of the remaining contamination, NCC placed a prohibition order on ConCen_Hall_1 which is still in force (See Annex VI). Entrance to the building is now controlled by locked doors to which only ConCenCo and NCC have keys. ConCenCo are not permitted to enter the Hall without prior written consent from NCC.

Table 1: Incident Summary

Initial Location of Incident	ConCen_Hall_1	
Substances Involved	2 litres of Niras and alcohol	
Weather at time of incident	The weather was dry and cool with a weak westerly wind.	
Timeline		
Time / Date	**Occurrence**	**Comment**
13:00:00 1/4/2012	Initial Event	During a religious conference. The Niras was released using a simple mechanical mechanism (the details of the actual method of release are confidential). This resulted in both liquid splashes and the release of vapour and spray.
13:10:00 01/04/2012	Emergency Services Arrival	Scene was attended by Ambulance (first to arrive) and fire services.
13:17:00 01/04/2012	Police Arrival	Police took charge of scene
13:23:00 01/04/2012	Complete evacuation of ConCen achieved and vent Systems isolated.	
03/04/2012	NCC place prohibition order on ConCen_Hall_1	All persons prohibited from entering Hall1 without written permission from NCC.
etc	Etc	Etc
30/04/2012	Police release of scene	Etc
Scope		
Areas affected	ConCen_Hall_1 and Foyer and ventilation systems	
No. People directly affected	300 people in the audience of whom 1 died and five treated	

3.3 Extent of Contamination

The extent of the contamination arising from the incident is presented in detail in Annex II and is summarised briefly below.

This information presents the best available data at the time of preparation of this Remediation Plan. Further sampling and investigation will be undertaken prior to and during the remediation works. Details of these arrangements will be presented in the Decommissioning Justification Document which will be submitted after approval of this Remediation Plan.

As noted above, Niras was released by some form of mechanical dispersion which resulted in liquid and spray being released – i.e. Niras was not directly vaporised by the incident. Annex II shows the most affected areas. Thus contamination has occurred through direct contact between the chemical and the fabric and structure of Hall 1. Because of the low volatility of the chemical vey little vapour will have been produced. Some chemical is known to have migrated through the foyer area during the emergency phase but this was treated directly by the HazMat team and has been confirmed as being of negligible residual risk.

The incident occurred in Hall 1 (outlined in Red on the layout shown in Annex II). The Hall is a modern facility (built in 2001) of traditional brick construction within a steel framework. It is served by a self contained ventilation system (now shut off and isolated). Access to the building is via a main entrance and foyer with emergency egress via fire doors as shown in Annex A1.1.

The HazMat team who attended the initial incident took samples of the contaminant and have since undertaken several monitoring and sampling exercises (details of the contaminant (Niras) are presented and discussed in Section 6). The available data from the HazMat Team (See Annex III) show that detectable levels of Niras remain in the hall area and to a much lesser extent in the Foyer.

The liquid Niras is believed to have soaked into the fabric of the furniture in the hall (chairs, curtains and carpets) as well as into the plaster wall finishes in places which were exposed to the spray and liquid. Vapour levels in the building are low.

No sampling has been conducted within the ventilation system, but there have been no significant detectable levels of Niras outside of the building (any transfer from contaminated staff to areas outside of the building have been naturally degraded by the weather).

4. Remediation Management Arrangements

ConCenCo Ltd intend to appoint a suitably qualified contractor to undertake the remediation works at ConCen_Hall_1 – such that the hall is then safe for re-occupation. Subsequent decoration and refurbishment will be undertaken by the in-house maintenance team.

The remediation contractors will report directly to ConCenCo Ltd's Technical Director (TD) who will be the single point of contact for all works relating to this incident (see Annex V). The Technical Director has been given full authority to act on behalf of ConCenCo Ltd in this matter.

The Technical Director will be responsible for co-ordinating with the remediation working group (see Section 10).

ConCenCo have appointed a specialist with knowledge of decontamination and of Niras as a technical advisor (TA) to the Technical Director. The TA's cv is attached in Annex VII.

The works will be undertaken in accordance with written method statements and risk assessments which will not be authorised for use until NCC have approved them. All works will be overseen by the TA.

NCC have similarly appointed one of their officers as responsible officer (RO) for the remediation works. The RO will liaise director with the TD and the TA as appropriate. The RO will have authority to stop works are any time.

The appointed contractor will be required to operate a certified qualify and environmental management system compliant to EN 9001 and 14001 respectively.

A waste disposal contractor will be appointed by ConCenCo in consultation with the appointed contractor.

The key roles and responsibilities agreed between ConCenCo and NCC are listed below;

Table 2: Key Roles for the Remediation Works

Role	Organisation	Responsibility / Authority
Technical Director (TD)	ConCenCo	Overall management of the remediation works for ConCen Co and co-ordination with NCC. To ensure co-ordination and transfer of information between ConCenCo, NCC and the Working group.
Technical Adviser (TA)	ConCenCo	Provision of expert advice and support to the TD and RO regarding control of hazards arising from Niras.
Responsible Officer	NCC	Overall management of the remediation works for NCC; approval of proposals from ConCenCo and co-ordination with NCC. To ensure co-ordination and transfer of information between ConCenCo, NCC and other interested government parties. To provide support to ConCenCo re public relations and press releases. To remove restrictions and covenants on the affected areas once remediation is complete.
Remediation Contractor	Tbc	To undertake remediation of the affected areas in accordance with written agreements with ConCenCo. To arrange for the removal and disposal of wastes
Etc		
etc		

4.1 Arrangements for Health, Safety, Quality and Environment

The ConCenCo Technical Director will have overall responsibility for all matters relating to Health, Safety and Environment (HSE). He will be advised by a Technical Adviser who will co-ordinate between ConCenCo, NCC and the appointed remediation contractor's representative.

ConCenCo will apply their existing ISO18001 approved safety management system to the works, expanded and modified as advised by the Technical Adviser.

All arrangements for HSE will be subject to written approval from NCC prior to implementation.

Details of the updated safety management system will be provided to NCC with the Remediation Justification.

Any pre-works required prior to the Remediation Justification will be undertaken in accordance with self contained written method statements subject to NCC approval.

4.2 Arrangements for Control of Subcontractors

ConCenCo's safety management system contains detailed guidance on the management of sub-contractors. These will be applied to this project. Only contractors who have management systems accredited to ISO 18001 and ISO14001 will be used on this project.

All subcontractor appointments will be subject to NCC's written approval.

4.3 Health Management

ConCenCo will arrange for health surveillance of its staff by an Appointed Doctor. Subcontractors will be required to demonstrate similar arrangements for their operatives.

5. Financial Arrangements

ConCenCo Ltd's primary insurers have agreed to cover costs for remediating the hall up to the policy limit of €2 Million (the primary tier insurance limit). ConCenCo Group have a further second tier limit of an additional €1 Million which could be called upon if necessary.

NCC have agreed to suspend charging of rates for the ConCen site for the duration of the incident (i.e. up to the time that Hall 1 is declared fit for re-occupation).

ConCenCo believe that they have sufficient resources to fund the proposed remediation works, which are not expected to exceed to primary insurance limit.

6. Hazardous Material Data

The Police and HazMat team have confirmed that the principal chemical used in the attack was Niras (see Annex III) mixed with pure alcohol to reduce its viscosity. It is suspected that the alcohol content has since evaporated and dissipated. Niras is a chemical warfare agent (CWA) which has previously been used by terrorists in attacks in public places.

Annex III shows that Niras is a colourless to slightly brown liquid with a slightly fruity smell. It is harmful to humans via skin contact and although it is only mildly volatile its vapour can also be harmful. It is lethal in large acute doses but leads to temporary disability in lower doses.

6.1   Hazards Arising from the Incident

As a result of the incident, Hall 1 is potentially hazardous to humans as there remain small amounts of the liquid and some vapour.

Skin contact with the liquid or inhalation of the vapour is potentially hazardous to anyone entering the hall but presents little hazard to those outside.

The hazards from Niras are well known and there are prescribed occupational exposure levels in legislation (see Annex III).   Specialised PPE and RPE is required for working with Niras contamination.   The chosen remediation contractor will be required to provide evidence of appropriate experience and training in the use of such equipment (see Forward Action Plan).

Niras naturally degrades into harmless components on exposure to warmth and moisture and can also be broken down more rapidly by exposure to mild hypochlorite bleach for example. Niras is completely degraded into harmless components by incineration at >600°C.

Bulk Niras contamination is readily detected using hand-held CWA detectors and/or detector papers.   Lower levels of contamination that may have migrated into surfaces require sample extraction and laboratory analysis.

6.2   Pre-existing Hazards

ConCen is a modern building complex, constructed in 2001. It does not contain any significant known hazardous materials; it has been subject to an asbestos survey which confirmed that none was present.

A review of the risk assessments for the building has not identified any hazardous material in the hall. No cleaning supplies are kept within the building.

7.   Proposed Remediation Goals and Standards

The purpose of this section is to present the remediation gaols and standards that will be adopted i.e. the extent to which it is proposed to remediate the affected areas and why.

7.1   Remediation Goals

ConCenCo Ltd wish to remediate Hall 1 such that it may again be used as part of the complex. ConCenCo Ltd does not wish to retain any of the furnishings or equipment in the building as we intended to completely refurnish with new equipment and furniture.

ConCenCo therefore wish to remediate the building so that it is safe for re-occupation and to dispose of affected furnishings and equipment.

7.2   Remediation Standards

Since it is intended that the remediation works will remove all contaminated (or potentially contaminated) materials that can be removed from the affected building, the remediation standards that are needed relate to the safe transport and disposal of these materials and to the levels of contamination that it will be acceptable to leave behind at the affected area so that it may be re-used as a public Hall.

In accordance with the guidance forwarded from NCC from Government Advisers (see Annex III), the level of acceptable residual contamination will be "non detectable using hand held CWA detectors " and "<1µg/kg as measured by mass spectrometry". A statistical sampling programme will be presented in the remediation justification document.

There are no extant values for the safe transport of materials contaminated with Niras; the chosen decontamination contractor will be asked to provide details of how they will provide safe transport of such materials such that any exposure of the transporters or the public will be assured (see Forward Action Plan) but it is anticipated that simple double wrapping in polythene will be sufficient.   These arrangements will be presented in the remediation justification document which will be approved by NCC in writing prior to any works.

8.   Proposed Remediation Plan and Programme

The proposed outline plan is to

1. seal the hall with polythene sheeting and to install a filtered ventilation system to maintain it a negative pressure. Exhausted air will be ducted through activated carbon filters to roof height and discharged. NCC have agreed to negotiate with the Environment Agency for a special discharge authorisation for this release (see Forward Action Plan).

2. treat any bulk contamination in-situ using BX90 (a proprietary decontamination agent)

3. dismantle and removable items (if necessary) and (subject to approval by NCC) double wrapped in heavy duty polythene and transport them by licensed waste carrier directly to a licensed waste site for incineration.

4. Treat any remaining surfaces with BX90.

5. Leave the room to dry for several days.

6. Obtain samples of any remaining surface materials and submit them for laboratory analysis

7. Remove any contaminated finishes

8. Apply a coat of non water based paint (e.g. acrylic paint) to all exposed surfaces prior to re-decoration and re-fitting.

A detailed description and justification of the works will be presented in the remediation justification document.

9.   Wastes

The expected waste arising from the remediation works are broadly as follows, but a detailed waste management plan will be developed as part of the remediation justification document. All wastes will be sent for secure incineration at a licensed site (to be confirmed – see FAP).

Item	Description	Approx Quantity
• Furniture (principally fabric coated seating)	Treated but potentially contaminated materials (note that metal frames will be decontaminated and kept if possible)	200 seats
• Carpet tiles	Treated but potentially contaminated nylon floor tiles	$100m^2$ (4mm thick)
• Plaster and building wastes	Treated but potentially contaminated scabbled surface finishes	Unknown, <1te
• Ceiling tiles	Fibre based tiles	$100m^2$ (4mm thick)
• Polythene sheeting	Fire retardant heavy duty polythene	Unknown
• Discarded PPE and RPE and other waste from the emergency response	Materials left by emergency services + contractor's materials	Unknown
• etc	etc	etc

10.   Stakeholder Involvement

ConCenCo have established a remediation working group which will be attended by employee representatives and union representatives. Members of the local community will be invited to attend meetings. ConCenCo will appoint an independent co-ordinator for the Group. ConCenCo intend to ensure that all decisions regarding remediation are referred to the working group for consent.

The working group have already endorsed this remediation plan subject to formal endorsement of the remediation justification document.

11. Conclusions

An incident has occurred at ConCenCo's conference facility in ConCen_Hall_1 which has left the area contaminated with residues of Niras, a chemical warfare agent.

A plan has been put forward which sets out at high level the arrangements that are proposed for the management and execution of the required remediation works in order to return the hall to a state where it will be safe for re-use.

Detailed substantiation of the proposed remediation works will be presented in a Remediation Justification Document, to be presented after approval of this plan.

Approval for re-occupation of the affected hall will be sought through the production of a remediation confirmation document which will also be submitted to NCC for approval.

12. Forward Action Plan

Guide: After completion of the previous sections, collate below any actions necessary to ensure that any identified gaps or requirements are addressed.

No	Action	Responsible Party	Timescale
1	Approve this Remediation Plan	NCC	3 weeks from submission
2	Prepare Remediation Justification Document	ConCenCo	6 weeks from approval of RP
3	Approve Remediation Justification Document	NCC	3 weeks from submission
4	Identify and agree a Waste Disposal Site	ConCenCo, NCC, Remediation Contractor	Prior to agreement of RJD
5	Obtain evidence of sub-contractor training in use of RPE and PPE for Niras.	ConCenCo	Prior to agreement of RJD
6	Confirm proposed transport arrangements for contaminated wastes	ConCenCo	Prior to agreement of RJD
7	Obtain agreement from Environment Agency for Aerial Discharges	NCC	Prior to agreement of RJD

13. Authority Sought

ConCenCo seeks permission from NCC to proceed to the detailed planning stage for the remediation of ConCen_Hall_1 as discussed above.   The approval by NCC of this Remediation Plan does not constitute authority to undertake any specific remediation works or activities. No remediation works will be undertaken until the Remediation Justification Document has been approved by NCC.

14. Literature

1.   Memorandum of Understanding between NCC and ConCo dated 1/2/2010

I   Annex I: Reference Documents Describing the Incident

Contents

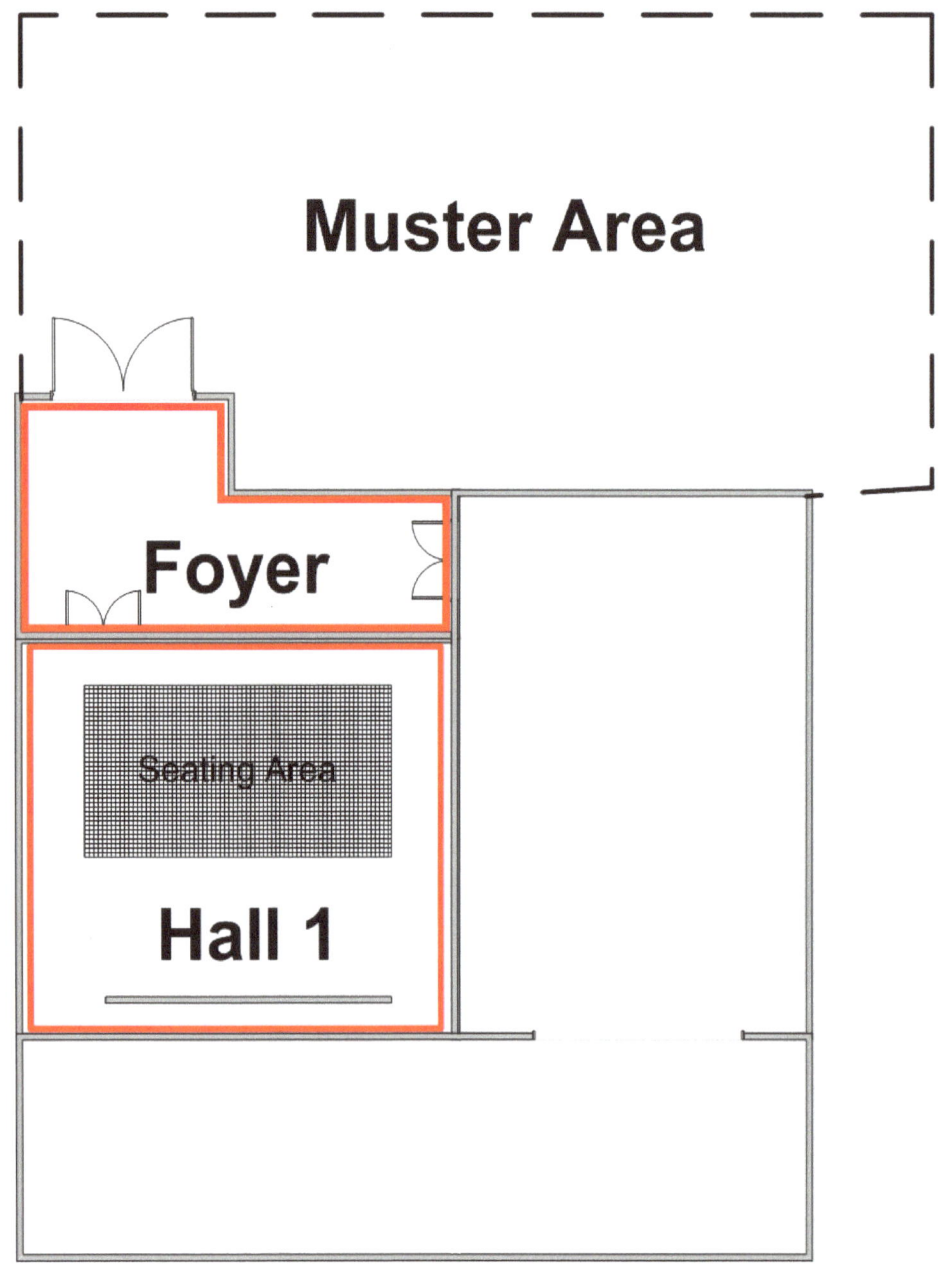

Figure A1.1 ConCen Layout Showing Hall1 and Contaminated Areas

18

II   Annex II: Extent of Contamination

Figure 1 above shows the areas where Niras concentration is highest (Red areas) and areas of detectable contamination (Orange areas).

Red areas indicate levels in excess of $1\mu g/m^2$

Orange Areas indicate areas $<1\mu g/m2$, $>.01\mu g/m^2$

III   Annex III: Details of Hazardous Materials

Name(s) of the Contaminating Substance(s)		C	B	R
Niras ($C_3H_4N_2OH$)		✓		

Existing Exposure Standards and Limits	Method of Measurement	Value		
Free Release / Re-Use	Mass Spectroscopy	<1µg/kg		
Transport	N/A	Double wrapped in heavy duty polythene and surface swabbed to contact standard (see FAP).		
Inhalation (Safe Working Level)	CWA Monitor	$0.1µg/m^3$		
Contact (safe exposure level)	Demin water swab over $1m^2$ and mass spec.	$0.1µg/cm^2$		
Known Health Hazards				
Hazard		Respiratory distress – failure in extreme cases		
Applicable Legislation		Directive EC/22/22/33.1		

IV   Annex IV: Legacy Hazardous Materials

No Legacy hazards have been identified.

There are no chemicals or other harmful substances within Hall 1.

V   Annex V: Organisation Charts and Contact Details

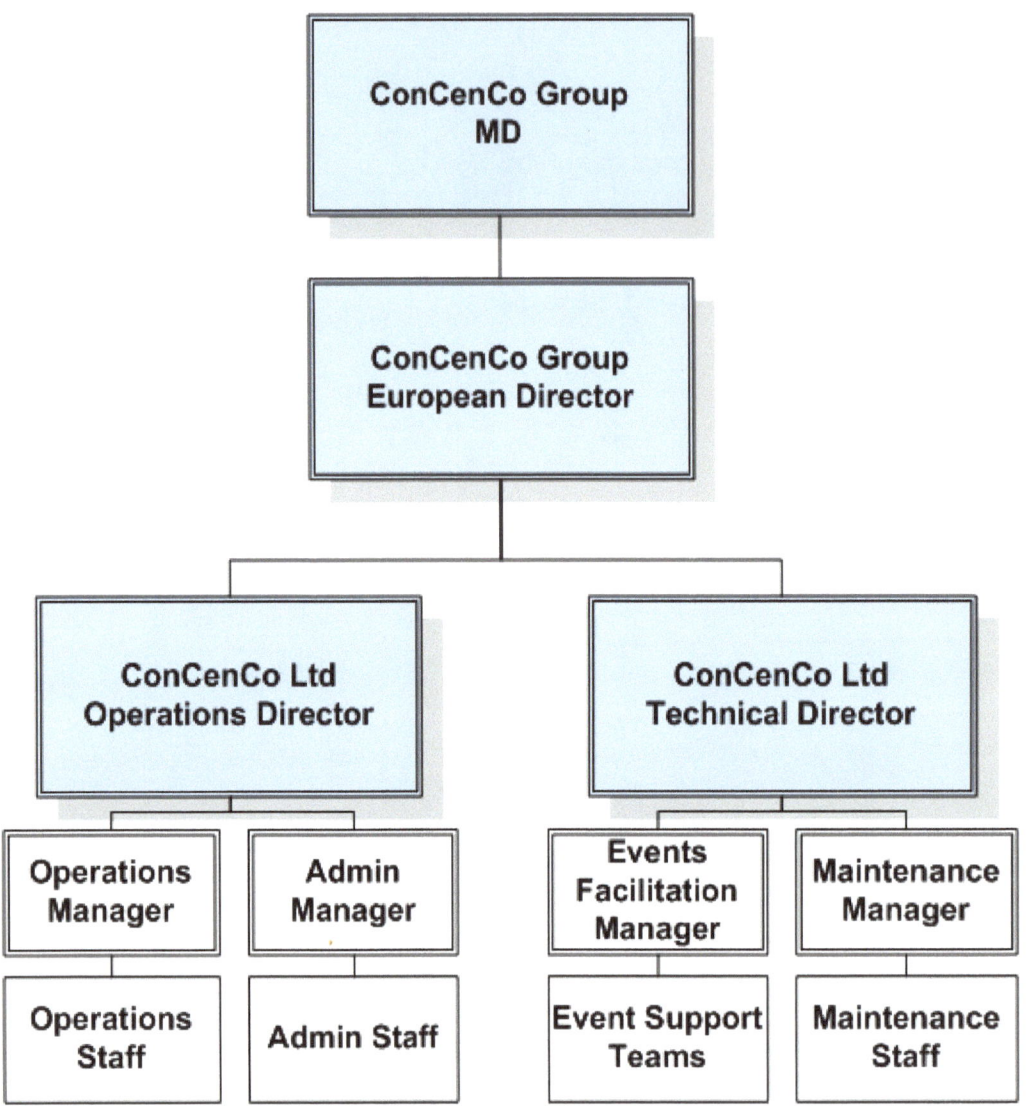

Figure A5.1 ConCenCo Organisation Chart

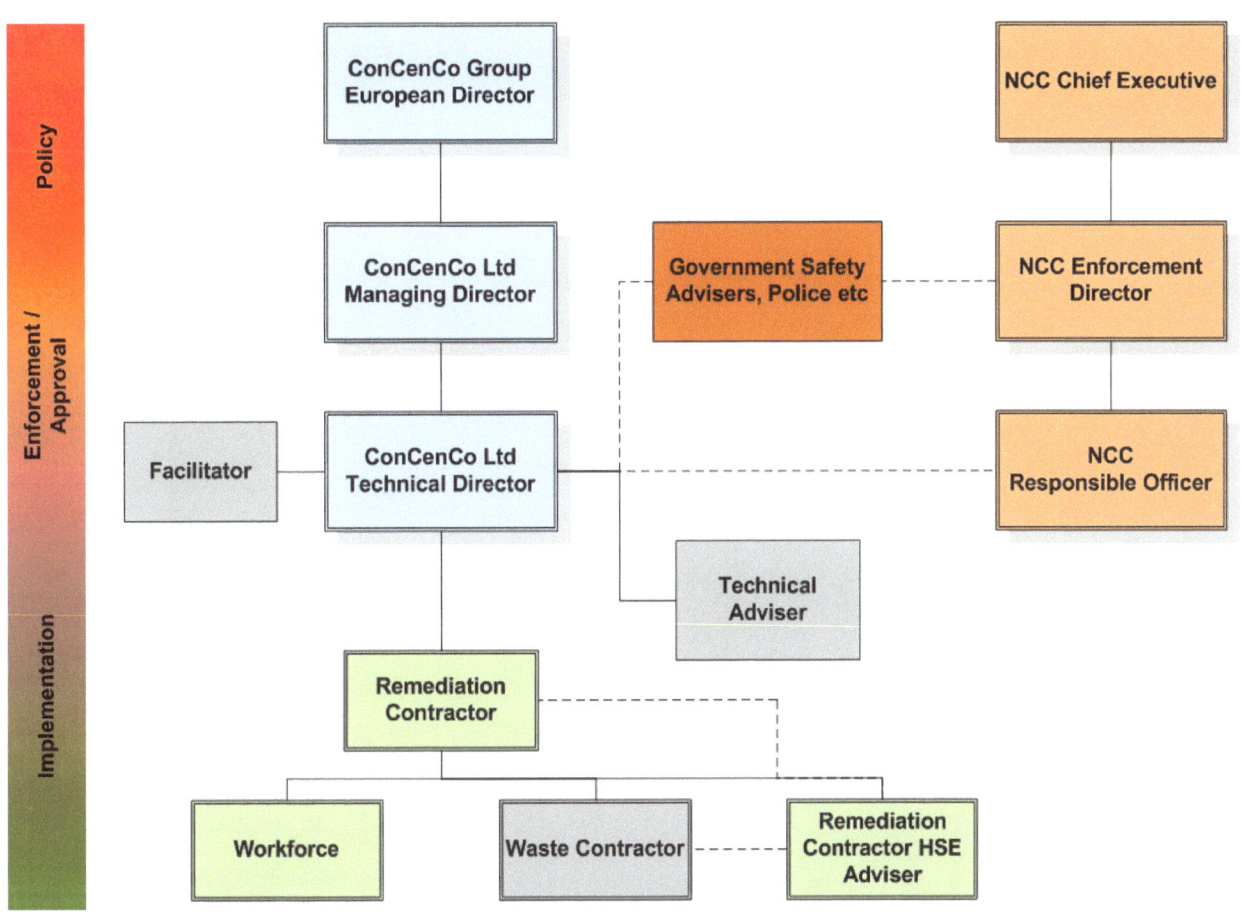

Figure A5.2: ConCenCo Remediation Management Organisation Chart

Table A5.1: Contact Details

Position	Name	e-mail	Tel No.
European Director (ConCen Group)	Francois Jones	f.j@concencogp.com	0033123456
Techncial Director (ConCenCo)	Frederik Kelly	Fred.kelly@concenco.com	00441234567
Etc	Etc	Etc	etc

VI   Annex VI: Details of Any Restrictions or Prohibitions

NCC have placed a prohibition on Hall 1 such that no persons may enter without prior written consent from NCC and ConCenCo other than emergency services and/or police.

A copy of the prohibition order is presented below;

NeuTown County Council

PROHIBITION ORDER

Emergency Restrictions Act 2007 – Section 22

NeuTown County Council (NCC) is satisfied that a Category 1 Hazard Exists at the commercial premises of ConCenCo Ltd at

Hall1, ConCenCo, Neutown, NT11 11A

The Council is further satisfied that the hazard presents an imminent risk to serious harm to the health or safety of any occupiers of these premises.

Under Section 22 of the Act the Council today imposes the following prohibitions;

    i)   The premises are not to be used for any purposes

    ii)   No person is to enter the premises without written permission from the Council's Director of Enforcement

    iii)   Any person entering the premises under a permission granted in accordance with item ii) does so in accordance with the conditions attached to that permission.

The schedule attached to this Order provides details of the hazard concerned and the remedial action which, if undertaken, would result in the Council revoking this order.

Dated 11/4/2007

Signed: ....

For Director of Enforcement

RO Anne Officer

Address for Enquiries: NCC
                                   1 High St
                                   Neutown, NT1 11A

VII   Annex VII: Miscellaneous

Contents

- CV for Technical Adviser
- Etc
- Etc
- Etc

REMEDIATION JUSTIFICATION FOR [*LOCATION*] STEP 2 OF 3

PREPARED BY [*ORGANISATION*]

APPROVED BY [*AUTHORITY*]

REVISION [??]

STEP 1: REMEDIATION PLAN

→STEP 2: REMEDIATION JUSTIFICATION

STEP 3: REMEDIATION CONFIRMATION

RECORD OF APPROVALS

ROLE	SIGNATURE		PRINT	DATE
AUTHOR				
APPROVED (ORGANISATION)				

APPROVED (AUTHORITY)			

Contents

Guidance to Authors for the Completion of this Remediation Justification - How to use this template

8. The following text strings, which appear in italicised text in this document, must be replaced with appropriate text throughout. Some of these replacements may be made prior to an incident but others will only be relevant once an incident has occurred.

Text String	Description
[Date]	The date of the incident which caused contamination of assets that now require remediation.
[Organisation]	The name of the organisation (company, entity etc) with responsibility for the assets and for ensuring their remediation to the satisfaction of the Authority.
[Location]	The Location of the Site(s) at which the contaminated assets are located
[Hazardous Material]	The name of the hazardous material which was principally involved in the incident and which has caused contamination of assets that now require remediation.
[Authority]	The name of the Authority with legal responsibility for ensuring that the remediation works are satisfactorily completed (e.g. a Local Authority, District Authority, State Office, Government Agency etc)

9. Text that appears at the start of a section like this example is overview guidance and should be deleted once the required text has been inserted.

10. Text that appears in normal black font is more detailed guidance as to the topics that should be addressed to satisfy the requirements of the section. This text should be deleted once the template is populated.

11. Text that appears {like this example} is example text that may be kept or modified as necessary by the Remediation Justification authors.

12. The Section headings should not be modified although additions are permitted.

13. This page may be deleted once the plan is complete.

14. A flow chart is shown overleaf which shows the process of completing this plan.

4

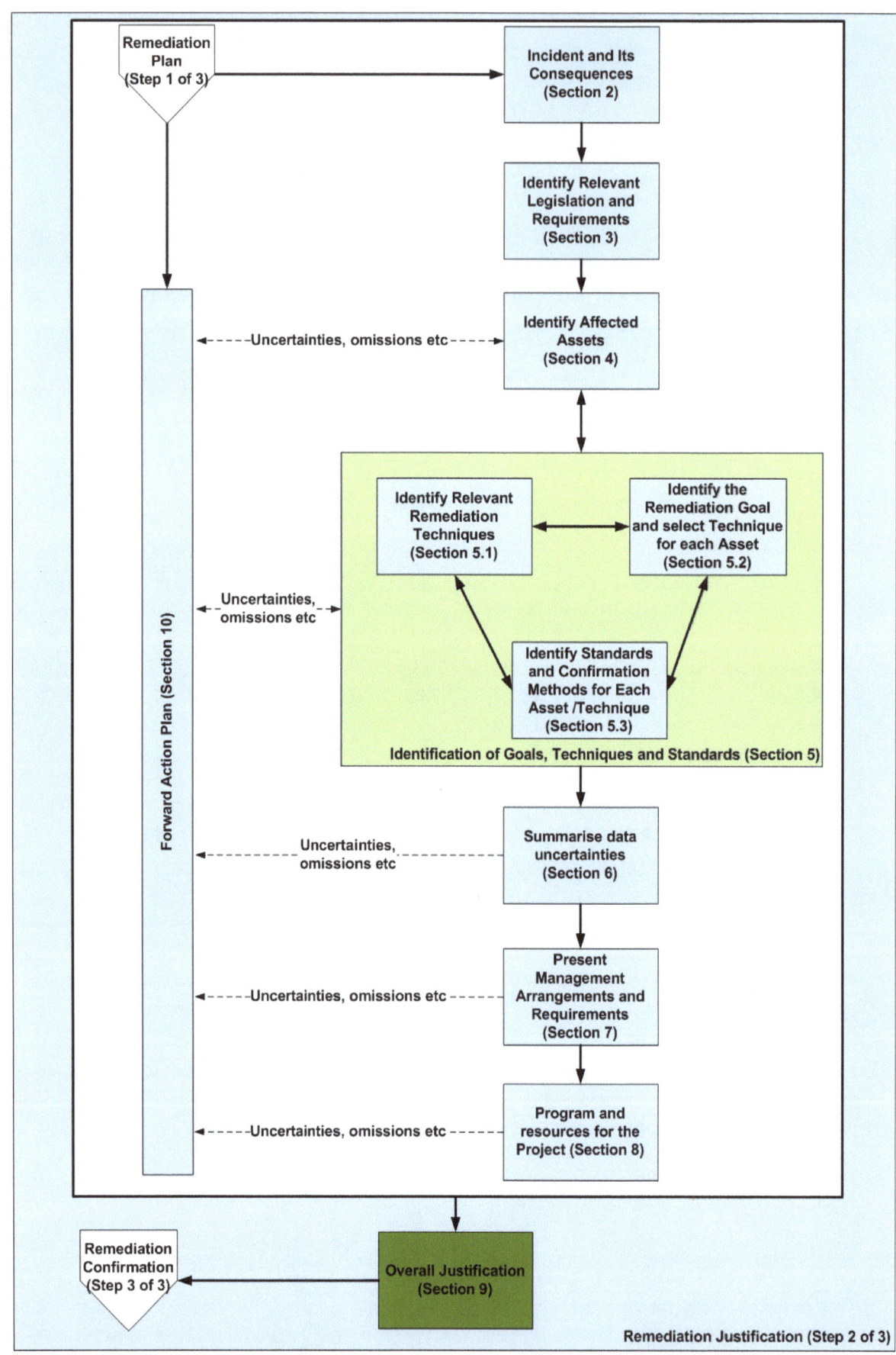

Flow Chart for Completion of the Remediation Justification

1. Introduction

The Key purpose of this document is to justify the selected remediation approach and its scope. Furthermore, it also presents details of the management arrangements and other controls that will be used to ensure that the proposed works will progress as planned and in accordance with relevant legislation. The approach is consistent with that used in high risk industries but has been simplified and streamlined.

The level of detail presented in the completed document should be commensurate with the complexity of the proposed tasks, their associated hazard potential and the hazard potential of the contaminants.

Some sections of this document will need input from third parties, such as those who were involved in the immediate response to the incident, or specialist bodies engaged to provide technical support and advice.

The main text should be kept as straightforward as possible with detailed technical information presented in the Annexes. In this way it is possible that the main document will be readable by and accessible to external third parties who may be interested but whom may not have detailed technical knowledge.

{On *[Date]*, some of the assets of *[Organisation]* at *[Location]* became contaminated as a result of an incident (the Incident) involving *[Hazardous Material]*. A Remediation Plan (Reference 1) which sets out the proposals for remediation of *[Location]* has been submitted to and approved by *[Authority]*. This Remediation Justification has been produced in line with the Strategic Agreement between *[Organisation]* and *[Authority]* as set out in Reference 2 to produce a series of Stage Submissions covering the required remediation works.

This Remediation Justification sets out exactly what is to be done and how it is to be executed and controlled. It therefore provides the detailed substantiation of the acceptability of the proposed works by reference to items such as the criteria to be used (e.g. remediation standards, legal framework), the actual remediation techniques to be applied and how they will be managed and controlled. The aim of this document is to present to the controlling responsible authority, sufficient evidence such that they are assured that the works to be undertaken will present an acceptable risk to all those concerned and that the works will comply with all relevant legislation.

This Remediation Justification seeks permission from *[Authority]* to undertake the works described here. Justification for termination of the works and release of the site from any special controls will be sought through the final stage submission called a Remediation Confirmation.

Any outstanding actions or uncertainties relating to this Remediation Justification are recorded in a Forward Action Plan which will be used to manage and monitor key actions that remain to be undertaken.

This Remediation Justification has been prepared using data provided by the *[Authority]* and the *[Organisation]*.}

2.   Details of the Incident and Its Consequences

This section provides an overview of the incident that has led to the need to produce this Remediation Justification.

Present a brief summary of the incident. It is appropriate to refer to the Remediation Plan for details, but a brief summary is required here to set the background for the remainder of this Remediation Justification.

In some cases, in the interval between preparation of the Remediation Plan and this Remediation Justification, further sampling and monitoring exercises may have been performed to confirm the nature and spread of contamination. Details of that exercise should be presented in Annex I and discussed here.

3.   Relevant Legislation

This section provides and overview of the legislation that relates to the hazardous materials involved in the incident. Legislation that relates to normal health and safety matters that would arise regardless of there having been an incident need not be listed.

Provide details of any legislation specific to the hazardous materials involved in the remediation works. Particular areas of concern will include aspects such as waste management (including disposal and transport), human exposure limits, transport regulations, chemical/biological weapons conventions and environmental regulations regarding discharges.

Provide details of any temporary relaxations that have been granted or which are necessary for the works to proceed. Any approval certificates or notifications which have been obtained or provided by others should be included in an Annex.

4.   Assets in need of Remediation

The purpose of this section is to explicitly identify the assets that need to be remediated.

Provide an overview of the assets covered by this Remediation Justification (refer to Annex II for detailed listings). The level of detail that the listing in the Annex should be broken down into depends upon the complexity of the proposed remediation tasks and the numbers of different types of assets. It may be necessary to re-visit the asset list and to break it down into further detail once Section 5 has been completed.

5. Goals, Remediation Techniques and Standards

The following sections should be used to set out i) the intended remediation goals (i.e. what is the intended destiny for each asset), ii) the remediation techniques (i.e. how those goals may be achieved) and iii) the remediation standards (i.e. the residual contamination/hazard levels which are applicable to the goals).

The selection of goals, techniques and standards is presented in the following sections as a linear process; it is recommended that as far as possible the sections and the associated tasks should be completed in the order presented. However, those completing this document should recognise that there will inevitably be some iteration before a consistent and acceptable set of goals, techniques and standards are arrived at.

Identification of the available remediation techniques is likely to require expert input.

5.1 Available Remediation Techniques

List the remediation techniques that may be suitable for the contaminated assets covered by this Remediation Justification. The purpose of the listing is to show that an appropriate range of techniques has been considered and that a balanced selection process has been used. For each technique the following items should be listed;

Item	Comment
Technique Name	A unique name
Brief Description	A brief description of the technique – e.g. spraying with decontamination foam X
Hazards Arising from Use / H&S Precautions needed	A list of any special hazards, e.g. decontamination foam X is mildly irritant to eyes – safety spectacles are required.
Pre-Treatment Required	List any special pre-treatment that may be required to ensure that the technique works properly - e.g. dismantling, size reduction, pressure washing, de-greasing etc.
Limitations	Identify any limitations of the technique – e.g. not suitable for absorbent surfaces.
Implications for Waste Management	Identify implications for waste management – e.g. foam is washed off items after use and is discharged to drains, after sampling
Reference Document	Identify a reference source for quoted information
General Suitability	Record overall comments regarding suitability of technique for this project

Note that it may be useful also to list some techniques that will not generally be applicable so that it is clear that they have been considered and dismissed appropriately, rather than being accidentally omitted. In some circumstances, the techniques to be used may be dictated by the relevant Authorities in charge of the incident. If so then this should be clearly identified.

The allocation of specific techniques to specific remediation goals for specific assets should be presented and justified in Section 5.2.

5.2 Detailed Remediation Goals and Technique Selection

For each contaminated asset it is necessary to define a remediation goal and to justify that selection. It is also necessary to select an associated remediation technique. The selection of the appropriate remediation goal and technique should be discussed and agreed with the Stakeholders.

Set out below, in detail, the remediation goals for each of the assets (listed in Section 4) and then present the selected Remediation Technique - note that each asset may have a different remediation goal, dependent upon its intended destiny. Starting with the decision to keep or dispose of the asset follow the decision tree in Figure A to select an appropriate goal. The standard set of 5 goals listed below should be used unless others are deemed more appropriate. In this latter case, justification for the new goals will be necessary

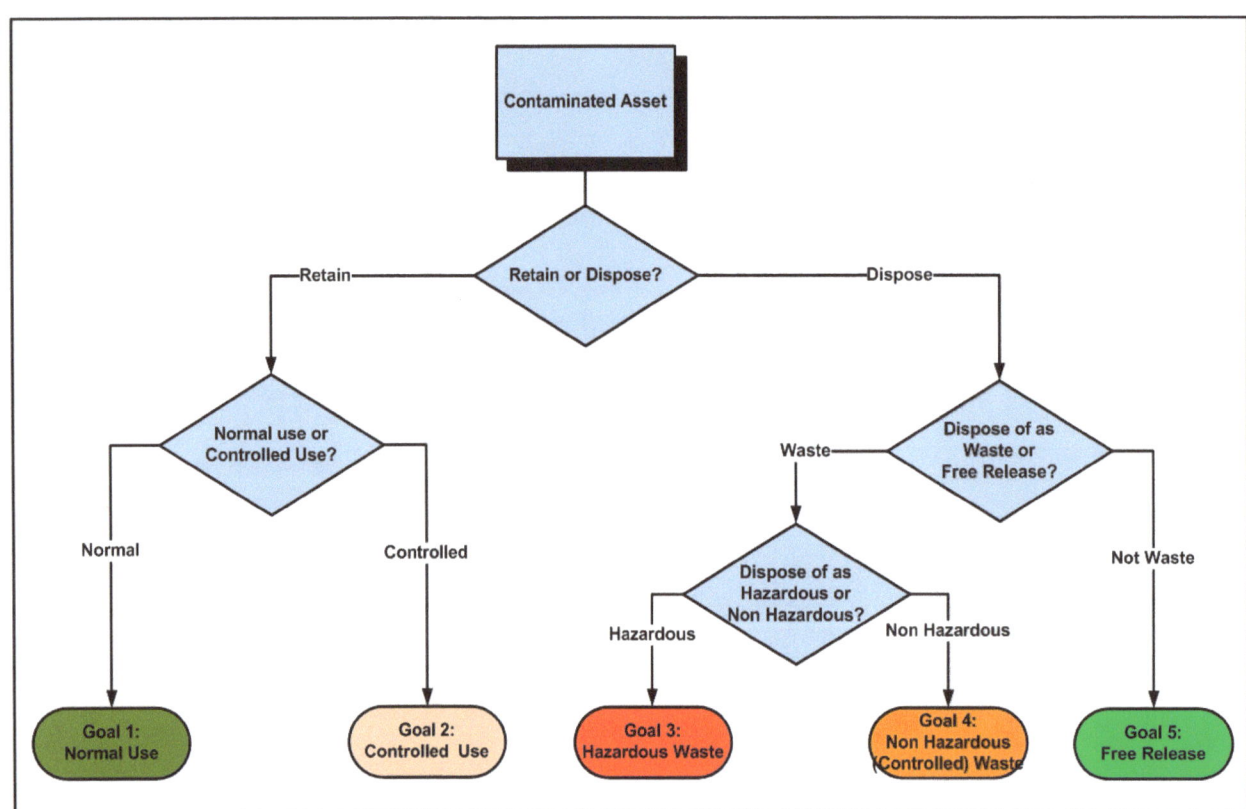

Figure A: Example Decision Tree for Goal Selection

➤ Goal 1 - assets that are to be re-used or retained by the organisation, without controls (normal use)

9

- Goal 2 – assets that are to be retained by the organisation under prescribed controls and safeguards (e.g. items may be kept in locked rooms or otherwise isolated from future use – controlled retention)

- Goal 3 - assets that are to be disposed of as hazardous wastes

- Goal 4 - assets that are to be disposed of (to controlled disposal sites) as non hazardous wastes and

- Goals 5 - assets that are to be released without any restrictions on their future use or destiny (i.e. free release).

Use the table below and drawings / photographs etc attached in Annexes to record and justify the proposed remediation goal and the associated remediation technique for each asset covered by this Remediation Justification.

Selection of Remediation Goals and Techniques

Asset ID	Asset Description	Applicable Pre-Decontamination Monitoring / Survey Results (if any)	Selected Remediation Goal	Justification for Selected Goal	Selected Remediation Technique(s) ID	Justification for Selected Technique for this Asset
Insert asset unique ID or range of IDs	Insert a brief description of the Asset and reference to drawing / photograph if available	Identify any existing monitoring / sampling data that relates to this asset	identify the selected remediation goal – from the list presented above	Insert a simple justification if possible. If the selected remediation goal is dependent upon the organisation providing ongoing protection or isolation from the hazard (e.g. Goal 2) then a more detailed justification for that selection should be presented.	Remediation Technique ID from Section 5.1	Insert a simple justification if possible. In complex cases the choice of appropriate remediation technique may not be clear-cut or straightforward and it may require more detailed assessment against multiple acceptance criteria. In this case separate assessments should be presented in Annexes and referred to from here. A simple example methodology is presented in Annex III. In other cases the selection may be very straightforward or there may be only one acceptable technique. In this case a simple explanation included here will be sufficient.

5.3 Remediation Standards and Confirmation

This section presents the allocation of remediation standards to assets. When the possible remediation standards were selected, agreed and documented during the production of the Remediation Plan then simply complete the table below. Where the standards were not available during the planning stage then it will also be necessary to provide detailed descriptions of the standards and their implementation in an Annex. If the Standards have been prescribed by an external Authority then this should be explicitly stated

Using the table below, for each of the Assets identified in Section 4, shows the remediation standard that will be applied during the works and the method that will be used to confirm that the required standard has been achieved.

Selection of Remediation Standards

Asset ID	Remediation Goal	Selected Remediation Standard	Reference Document / Source	Method of Confirming that the Standard had been met
Insert asset unique ID or range of IDs	Identify the selected remediation goal – from the list presented at 5.2 above.	If these have previously been listed in the Remediation Plan then insert the name of the Standard here, otherwise insert a Reference to the appropriate standard in the Annex.	Identify the source of the standard – e.g. existing legislation, directive from Authority, technical paper etc	Identify the method / technique by which confirmation of the standard will be confirmed, e.g. surface swab, vapour analysis, visual inspection etc.

6.   Uncertainties

This section is used to highlight any uncertainties or unknowns that may affect the works.

Present any limitations or uncertainties with regards to the data presented in the previous sections. In particular highlight any uncertainties regarding the spread of contamination or the hazards associated with it.

7. Waste

This section presents the routes to waste disposal.

7.1 Waste Types and Quantities

Identify the various waste streams that will be produced by the remediation works. For each waste stream present the following details;

Item	Explanation / Example
Waste Description	General description of the waste (e.g. contaminated furniture, building rubble, liquids etc)
Hazardous Content	Where the waste is hazardous waste (as defined by the Hazardous Waste Catalogue identify the hazardous components.
Estimated Quantity	m3, kg etc
Intended Waste Route	Landfill, incineration, hazardous landfill
Appropriate Hazardous Waste Catalogue Code	e.g. 16 09 04 oxidising substances, not otherwise specified
Proposed packaging arrangements	e.g. drums, bags, skips[18]

In instances where the wastes will not be sent directly for disposal, such as when final disposal arrangements are still under negotiation or where temporary storage is required because of the logistics of waste handling then identify how / where wastes will be stored pending disposal. Temporary storage of wastes at the site of production or other site operated by the producer is permitted under EU law but storage off-site is subject to additional controls. In the event that significant quantities of waste are to be stored then further advice regarding the suitability of storage facilities can be found in PRACTICE Deliverable D5.11 "Criteria for the selection of temporary waste storage facilities".

Where the final disposal facility is known[19], list the following

- The identity of the Waste Disposal Facilities that will be used for non hazardous waste

    Name:

    Address:

    Contact Person and Phone Number:

---

[18] Packaging should ensure that the contaminants will not escape under reasonably foreseeable conditions
[19] If the name of the precise facility is not known it will be appropriate to substantiate that suitable facilities do exist.

Confirmation that they will accept the waste (include confirmation letters etc in an Annex)

- The identity of the Waste Disposal Facilities that will be used for hazardous waste

   Name:

   Address:

   Contact Person and Phone Number:

   Confirmation that they will accept the waste (include confirmation letters etc in an Annex)

   The physical form that the waste is to be submitted in (e.g. bags, drums, solidified etc)

It may be appropriate to witness the transport and destruction of some wastes to provide assurance that they have been appropriately handled.

7.2 Segregation Plans

Identify how the different types of wastes will be kept separate to avoid cross contamination and minimise potential for wastes to be incorrectly sent to the wrong disposal route (e.g. separate storage areas, different coloured waste bags etc).

8. Management Arrangements

This section presents the arrangements that will be in place for ensuring that the proposed works will be undertaken safely and in conformance with all relevant safety legislation and requirements. This section does NOT replace risk assessments and method statements – although these should be referred to here. Although these arrangements have already been presented in summary in the Remediation Plan then they should be expanded upon here in greater detail. Any deviations or changes since preparation of the Remediation Plan should be explicitly highlighted.

The following topics should be addressed;

8.1 Training and Accreditation

List any special requirements for training and accreditation of the workforce (e.g. accreditation to work with hazardous materials, ionising radiations, biological agents etc). The arrangements for recording and monitoring such training requirements should also be noted – e.g. these may be recorded in method statements or risk assessments.

8.2 Monitoring (auditing, Inspections etc)

Present the arrangements for monitoring the safety performance of the workforce, for example through audits and inspections. If third party specialists are to be utilised to perform this then record their details (e.g. name, qualifications and contacts details).

8.3 Emergency Arrangements

Given the nature of some of the incidents and hazardous materials that could be covered by this plan, it is essential to clearly note here the arrangements for emergency situations or unexpected occurrences and the arrangements for notification of the appropriate authorities. In particular it may be appropriate to note that the local hospital has been notified of the type of work being undertaken and of the potential for casualties who have been exposed to the hazardous materials covered by these works

8.4 Arrangements for Health Management and dose monitoring

Present any arrangements that may be required for the monitoring and recording of the health of those potentially exposed to the contaminants (e.g. medical checks / surveys).

Where there is a requirement for on-going monitoring of the exposure of the workforce or others that may be exposed, then the arrangements for that monitoring should be noted here. In particular, the frequency and type of monitoring should be recorded. Wherever practicable an exposure (dose) target should be set for the workforce and the public. Actions to be taken upon reaching the prescribed targets should be highlighted. It is likely that these targets will be set by the Authority or its scientific advisers.

For the public and other external stakeholders (i.e. those not directly involved with the remediation works) requirements for monitoring are likely to be advised by the Authority or their technical advisers. For remediation workers and other employees / contractors of the Organisation these arrangements will be the responsibility of the Organisation but will likely also be advised by the Authority and their advisers.

If laboratory analysis is required then the details of the laboratory (name, address, contact details) and its accreditation status (relevant QA accreditation or sample analysis accreditation) should also be recorded.

8.5 Interim Reporting and Performance Monitoring

Present arrangements for providing interim reports and monitoring data to the Authority –e.g. the Authority may wish to have weekly progress reports or meetings.

8.6 Approvals and Hold Points[20]

Present any hold points and approval points agreed with the Authority.

8.7 Approvals of Variations

It is likely that the scope of works presented here will change as the work proceeds and knowledge and experience is gained. Where such variations occur it is good practice to have a simple method for categorising them according to their significance (in terms of safety and cost) and an agreed method for approving the variations. Present here the arrangements agreed between the Organisation and the Authority[21] for the management of variation to the scope of work described here.

8.8 Arrangements for Control of Subcontractors

Present the arrangements specifically relating to the control of sub-contractors – i.e. demonstrate how overall responsibility and control for the project rests with the Organisation, even though

---

[20] Points in the programme of work where, for example, the Authority has stated that they will require formal notification and their approval must be obtained before proceeding further. Examples could include trials on new or novel remediation techniques or completion of preparatory works prior to removal of a significant hazard.
[21] There will need to be a consistent agreement between the Organisation and it sub-contractors. See for example Managing Reality – Book 4 Managing Change, Thomas Telford Ltd 2005, ISBN 0 727 3395 8.

specialist contractors may be used for some of the actual works, As above, reference to existing management systems is acceptable as long as these are shown to be relevant to the types of works covered by this Remediation Justification.

Note: The Authority may wish to formally approve the use of any sub-contracting organisations (by name) prior to their engagement. This may be included within any advance framework agreement with prime contractors.

Arrangements for the production and approval of method statements should be presented. The Authority may wish to approve or note all method statements prior to the commencement of work. Similarly the responsibility for the production and approval of risk assessments should be noted.

The production and use of Quality Plans to list and record approvals of documents is recommended.

8.9   Security

Identify the methods by which security of the site and in particular that of any hazardous storage areas will be maintained during the works – e.g. the work areas and waste storage areas may require security patrols or CCTV supervision during silent hours because of the hazards they present and the nature of the contaminants that may be present (see Waste Management below).

8.10 Environmental Management

Summarise the plans for the management of environmental impacts from the proposed remediation works. For example, if the works are expected to be noisy or to produce dusts or other 'fugitive' emissions then the management methods use to mitigate these should be recorded briefly here.

Example impacts that could be considered are;

➤ Time Delays	Delay involved in the preparation and execution of remedial works / additional environmental damage estimated to arise due to necessary delays (e.g. for analysis, groundwater monitoring, etc).
	Also time taken to achieve anticipated remediation goals.
	The time taken to achieve remediation goals may affect whether a regulator will agree to a particular scheme. For instance, biological remediation can take months or even years to achieve.
➤ Residual environmental hazards	Level of acceptable residual hazards anticipated after remediation, based on "fit for purpose" considerations and expected environmental impacts. (i.e. post-remediation site sensitivity according to use). This includes social factors such as public access and proximity of residential areas as well as impacts on flora and fauna.
	This could be use-dependent rather than site-dependent. For example, more stringent remediation targets may be necessary

for items destined for public use but the existing levels could be acceptable for industrial use due to different exposure profiles.

- ➤ Public Health

  Level of risk to health involved in residual hazard. This includes occupational risk, public access, groundwater contamination risk and drainage.

- ➤ Waste Management and Resources Burden

  Level of waste management and resources required to perform operation, including types of wastes and their disposal routes, resource consumption and traffic.

- ➤ Aesthetics

  Any temporary or permanent impact on the publicly visible structures or areas – these will often be covered by planning and development regulations.

- ➤ Fugitive emissions

  e.g. noise, dust, vapour, steam

In the event that the project is large and complex then more significant environmental impacts may occur which may need to be addressed. In this instance a separate and more detailed environmental impact assessment will be required..   In that event, the separate assessment should be referenced here and the conclusions from that assessment should be summarised.

9.   Programme and Resource

This sections shows that the works have been appropriately planned and that adequate resources have been provided.

Provide an outline programme for the works and identify the resources that will be utilised (in terms of numbers and types of workers). Identify where the resources will be provided from – e.g. in house or sub-contract.

The programme should demonstrate a logical sequence to the proposed works. The programme sequence is likely to be driven by logistical considerations such as accessibility, but it should also recognise that where there is uncertainty regarding the approach to be taken then it is good practice to start the decontamination work with the less hazardous areas so that experience may be gained without undue risk. Once uncertainty has been addressed then, where practicable, areas which pose the highest immediate risk should be addressed first.

10. Overall Justification

This section summarises the reasons why the remediation works are considered to be justified – i.e. that they are safe, reasonable and appropriate.

Present a high level statement of the reasons why the works are considered to be justified, for example:

"Following an incident at [Site] on [Date], some of the assets of [Organisation] have become contaminated. The Remediation Plan, which was prepared and submitted, set out the remediation goals of [Organisation] and outlined the high level arrangements for the control of works that we now wish to undertake. The Remediation Plan has since been approved by [Authority].

This Remediation Justification document presented here, further develops the arrangements presented in the Remediation Plan and in particular it shows that [Organisation] has

➤ properly planned the works and can resource them

➤ selected appropriate remediation goals for the works with regard for potential future use of the contaminated assets

➤ selected appropriate remediation techniques –with due consideration for factors such as safety, environmental impact, cost and practicality

➤ selected an appropriate end-point for the works with due regard to future use of the site

➤ provided adequate arrangements for the management and control of the works with due consideration for safety, need for external approvals, hold points, use of sub-contractors and waste management issues.

➤ Identified any further actions that are required in order to resolve outstanding issues or uncertainties."

11. Forward Action Plan

The Forward Action Plan, presented here, is used to identify outstanding actions and items requiring further analysis or assessment in order to progress the project through to completion.

After completion of the previous sections use the Table below to collate any actions necessary to ensure that any identified gaps or requirements are addressed. Show progress of actions raised at the Remediation Plan stage.

No	Action	Responsible Party	Timescale	Progress
1	Approve this Remediation Plan	[Authority]		
2	Prepare Remediation Justification Document	[Organisation]		
3	Approve Remediation Justification Document	[Authority]		
4	Etc			

12. Authority Sought

This section clearly identifies the scope of permission sought from external parties.

Clearly set out the exact scope of permission that is being sought and from whom. Identify any hold-points or further permissions that may be sought —e.g. "Permission is sought from the Local Authority to commence remediation of the Site as set out above. Permission to re-open the site for normal use will be sought separately after completion of a Remediation Confirmation document."

13. Literature

Sampling Results

Sample ID Number	Sample Location*	Date Obtained	Analyte tested for	Area Sampled	Limit of Detection	Results Obtained	Reference Document *

Monitoring results

Monitoring Result Id	Monitoring Location *	Equipment Used	Area Monitored	Date Monitored	Limit of Detection	Results Obtained	Reference Document *

*Photographs and drawings should be used to show the locations of samples obtained.

** Reference to detailed survey report or laboratory report (attach if possible)

II   Annex II: Assets to be Remediated

Identify the assets that have been contaminated and which are in need of remediation. Organisations may also wish to note the value of the assets if this has a significant bearing on the choice of whether to remediate or dispose.

Asset ID	Sample / Monitoring Reference (if available)	Description	Drawing / Photo Reference	Location	Principal Materials of Construction	Condition
		{e.g. machine, chair, floor, wall etc}			{e.g. metallic, wood and fabric, concrete, plaster board, etc}	{e.g. pitted, as new, newly painted, local areas of damage, etc}

III   Annex III: Detailed Selection of Remediation Methods, a simplified approach.

Use this Annex where it is necessary to present a more detailed comparison of different remediation techniques for particular assets or asset types. It may be useful to complete this comparison in conjunction with a Stakeholder Group, under scientific and technical guidance. Some example criteria and a simple comparison method are presented. More complex ranking methods and weighted summing techniques are also available, such as those listed in "Department for Communities and Local Government (DCLG) 2009, *Multi-criteria analysis: a manual, Department for Communities and Local Government*, Eland House, Bressenden Place, London, SW1E 5DU" and ".Müller-Herbers, S (2007), *Methoden zur Beurteilung von Varianten, Fakultät Architektur und Stadtplanung*, Institut für Grundlagen der Planung" and "Lehoux N, Vallée P (2004), *Analyse Multi-critères*, available at

http://www.performance-publique.budget.gouv.fr/fileadmin/medias/documents/performance/controle-gestion/Qualite_et_controle_de_gestion/Analyse_multicriteres/1_Multi_criteres2004.pdf.

Expert advice should be sought in the use of more complex techniques.

The proposed criteria to be used to compare the available remediation methods are;

Table AIII.1: Comparison Criteria

Reference	Criterion	Explanation
C1	Efficacy	Proven ability to remediate against the hazard (for the type of asset considered) – i.e. has the technique been used previously for the same type of hazard and asset?
C2	Cost	Cost of implementation and cost of remedying any impact from the technique (e.g. the technique may damage the asset such that repair costs will also be incurred).
C3	Safety Implications	Hazards associated with use of the technique (inc implications for future handling of the asset)
C4	Damage	Potential for damage to the item being remediated
C5	Ease of Application	e.g. Some techniques may require significant preparatory work on contaminated assets – for example, dismantling may be required in order for the technique to be applied to complex items and surface shapes, whilst other techniques – like those that involve fuming or exposure to vapours may need less preparation.
C6	Environment	Impact on the environment (e.g. from emissions, secondary wastes, waste disposal, use of resources – water, air, power etc - noise issues etc).

Table AIII.2: Comparison of Methods

Asset						
Criteria	Method1	Comment	Method2	Comment	Method3	Comment
C1	O	Etc	X	Etc	✓	Etc
C2	X	etc	✓	etc	X	etc
C3	X	etc	O	etc	✓	etc
C4	✓	etc	O	etc	X	etc
C5	✓		✓		X	
C6	✓	etc	✓	etc	O	etc
Overall Score [22](Sum)	+1		+2		-1	

X   Method does not perform well against the criterion (Score -1)

O   Method is neither good nor bad against the criterion (Score 0)

✓   Method performs well against the criterion (Score +1)

---

[22] Higher scores indicate a higher preference or rank for the method.

REMEDIATION CONFIRMATION FOR [LOCATION]

PREPARED BY [ORGANISATION]

APPROVED BY [AUTHORITY]

REVISION [??]

STEP 1: REMEDIATION PLAN

STEP 2: REMEDIATION JUSTIFICATION

→ STEP 3: REMEDIATION CONFIRMATION

RECORD OF APPROVALS

ROLE	SIGNATURE	PRINT	DATE
AUTHOR			
APPROVED (ORGANISATION)			
APPROVED (AUTHORITY)			

Contents

Guidance to Authors for the Completion of this Remediation Justification - How to use this template

15. The following text strings, which appear in italicised text in this document, must be replaced with appropriate text throughout. Some of these replacements may be made prior to an incident but others will only be relevant once an incident has occurred.

Text String	Description
[Date]	The date of the incident which caused contamination of assets that now require remediation.
[Organisation]	The name of the organisation (company, entity etc) with responsibility for the assets and for ensuring their remediation to the satisfaction of the Authority.
[Location]	The Location of the Site(s) at which the contaminated assets are located
[Hazardous Material]	The name of the hazardous material which was principally involved in the incident and which has caused contamination of assets that now require remediation.
[Authority]	The name of the Authority with legal responsibility for ensuring that the remediation works are satisfactorily completed (e.g. a Local Authority, District Authority, State Office, Government Agency etc)

16. Text that appears at the start of a section like this example is overview guidance and should be deleted once the required text has been inserted.

17. Text that appears in normal black font is more detailed guidance as to the topics that should be addressed to address to satisfy the requirements of the section. This text should be deleted once the template has been populated.

18. Text that appears {like this example} is example text that may be kept or modified as necessary by the Remediation Confirmation authors.

19. The Section headings should not be modified although additions are permitted.

20. This page may be deleted once the plan is complete.

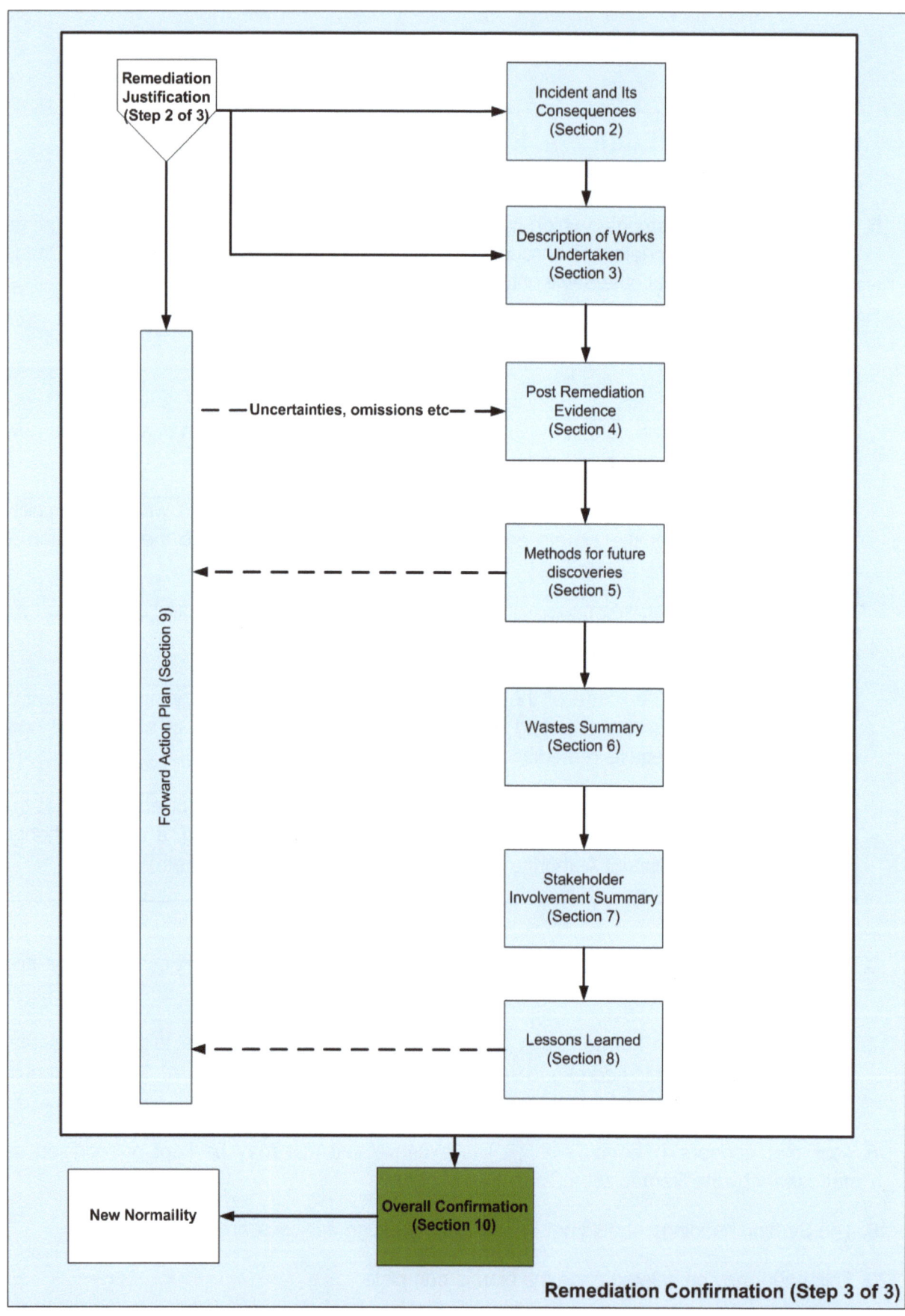

Flow Chart for Completion of the Remediation Confirmation

4

1. Introduction

This Remediation Confirmation template is intended to provide a vehicle for the presentation of the results of the remediation works presented in the Remediation Plan and the Remediation Justification. Its completion will also provide the vehicle for the controlling Authority to show that the works have been completed to its satisfaction and that the affected Site is now safe for re-use.

The main text should be kept as straightforward as possible with detailed technical information presented in the Annexes. In this way it is possible that the main document will be readable by and accessible to external third parties who may be interested but whom may not have detailed technical knowledge.

{On *[Date]*, some of the assets of *[Organisation]* at *[Location]* became contaminated as a result of an incident (the Incident) involving *[Hazardous Material]*. A Remediation Plan (Reference 1) which set out the proposals for remediation of *[Location]* was submitted to and approved by *[Authority]*. The method by which that Plan was subsequently implemented was set out in a Remediation Justification (Reference 2) which was submitted to and approved by [Authority].

This Remediation Confirmation sets out what was done, what wastes were produced, what the condition of the site is now and what lessons were learned that may be of use to others in the future. The aim of this document is to present to the controlling responsible authority, sufficient evidence such that they are assured that the works can be declared as complete and released from their control.

This Remediation Confirmation seeks permission from *[Authority]* to [insert summary of the permission sought]

Any outstanding actions or uncertainties relating to this Remediation Confirmation are recorded in a Forward Action Plan which will be used to manage and monitor key actions that remain to be undertaken.}

2.   Details of the Incident and Its Consequences

Present a brief summary of the incident. It is appropriate to refer to the Remediation Plan for details, but a brief summary is required here to set the background for the remainder of this Remediation Confirmation.

3.   Details of Works Undertaken

Present a summary of what work was actually performed – this may be slightly different from what was actually planned and proposed by the Remediation Plan and the Remediation Justification. Any such differences should be explicitly highlighted and justified. Where separate authorisations were obtained for these variations then these should be attached in the Annexes and referenced here.

Provide a summary of the remediation processes / techniques that were applied.

Provide before and after photographs where practicable and drawings/plans showing the locations of the remediation works that were undertaken.

3.1   End State Description

Present a summary of the end-state of the Site in terms of what has been achieved and whether there are any residual hazards present (e.g. all detectable contamination by "X" has been removed from the Site. However, it is suspected that some contaminants remain, below detection levels, in the ventilation system). Identify any long term monitoring that may still be required (e.g. ongoing monitoring of the ventilation system by a contractor will be used to ensure that levels of "X" remain below detectable limits).

3.2   Incidents

Present a summary of any incidents that occurred during the remediation that are worthy of note. It is especially important to highlight any unusual occurrences or incidents that may have led to further spread of contamination – e.g. any spills or breaches of containment.

3.3   Records

Identify where the project records are kept and the proposed period of retention.

4.   Details of Post Remediation Evidence

Detailed evidence should be provided in Annexes to this report. Present a summary of that evidence here to support the claim that the remediation works have been effective. Clearly identify what has been done, what the detection limits were and how the results compare with the proposed remediation goals and standards presented in the Remediation Justification. Where external laboratories are used it is important to note their name and their accreditations as well as detection limits – include laboratory reports in an Annex.

Examples of evidence include;

- Analysis of samples (e.g. analysis of swabs of assets)

- Monitoring of areas (e.g. air monitoring to detect the presence of airborne contaminants)

- Visual inspections (e.g. before / after photographs)

- Witness strips[23]

It is good practice to ensure that those obtaining the post remediation evidence are independent of those who performed the remediation works. The Authority may also wish to appoint its own surveyors and analysts to obtain further confirmatory evidence. If that is evidence available at the time of production of this report then it should also be presented here.

As some methods of sampling, such as taking swabs, removes some contamination, it is important to maintain a careful log of sample locations so that the same area is not sampled twice.

5. Method of dealing with any future related contamination that may be discovered

Provide a summary of any proposed methods for dealing with the discovery of any unsuspected residual contamination in the future.

6. Wastes

This section is used to provide information on the wastes that were produced and evidence that they have been appropriately disposed of or destroyed.

Provide summaries of the following data

- volumes (and/or masses) of wastes produced by the works, segregated by type (hazardous / non hazardous etc) and form (solid, liquid etc).

- evidence (supported by detailed evidence in an Annex), that the wastes have all been disposed off to appropriately licensed waste disposal / destruction sites e.g. provide waste transfer notes, waste contractors registration numbers etc, waste destruction certificates.

---

[23] A witness strip is a piece of material that has been coated with a sample of the contaminant (or a surrogate for the contaminant) that is being removed by the remediation. They are placed in strategic locations prior to the works and are removed and tested after the works. If the contaminant has been removed or destroyed on the witness strip then this is taken as evidence that the remediation works have been effective.

- evidence of any witnessing[24] of waste destruction/disposal that was undertaken

7.   Stakeholder Involvement

> This section is used to show that any appropriate Stakeholder engagement programme has been undertaken.

Present a summary of any workshops or other stakeholder engagement processes that have been undertaken and evidence that they have indicated their acceptance of the completed remediation works.

8.   Lessons learned

> This section is used to identify anything that was learned during the works that may be beneficial to others who may have to undertake similar works in the future.

Provide a summary of any difficulties encountered or incidents that occurred. Where possible provide details of any changes to process and procedures that were required to remedy these.

Specifically identify where any of the remediation techniques proved more or less effective than was envisioned at the Remediation Justification and Remediation Plan stages.

9.   Forward Action Plan

> The Forward Action Plan, presented here, is used to identify outstanding actions and items requiring completion.

At this stage any outstanding actions should be of a trivial nature and all actions previously identified should have been completed. Forward actions should be listed as in the following table. Include statements on the progress of actions identified in the Remediation Justification.

Forward Action Plan

No	Action	Responsible Party	Timescale	Progress

---

[24] It is good practice to follow sample waste consignment to ensure that they were delivered to the facility stated by the transporter.

10.   Overall Confirmation

Provide a written overall confirmation that it is appropriate to re-open and operate under a new normality, e.g.

{The remediation of the Site has been planned and executed in accordance with a Remediation Plan and a Remediation Justification. Stakeholder and official acceptance of these has been obtained throughout the project. The evidence presented here shows that the assets that were contaminated have now either been decontaminated to the extent that they are now safe for their proposed use or that they have been safely disposed of to an authorised waste disposal site.

Suitable arrangements have been made for dealing with any future discovery of contaminants associated with this incident although there is no evidence that such contamination persists at the site.

All records generated during the works will be retained by the organisation for a suitable period and will be available for inspection.}

11.   Authority Sought

This section clearly identifies the scope of permission sought from external parties.

Clearly set out the exact scope of permission that is being sought and from whom. Identify any hold-points or further permissions that may be sought –e.g. "Permission is sought from the Local Authority to fully re-open the Site for normal use"

12.   Literature

1. The Remediation Plan

2. The Remediation Justification

Sampling Results

Sample ID Number	Sample Location*	Date Obtained	Analyte tested for	Area Sampled	Limit of Detection	Results Obtained	Reference Document*

Monitoring results

Id	Monitoring Location*	Equipment Used	Area Monitored	Date Monitored	Limit of Detection	Results Obtained	Reference Document**

*Photographs and drawings should be used to show the locations of samples obtained.

** Reference to detailed survey report or laboratory report (attach if possible)

www.ingramcontent.com/pod-product-compliance
Lightning Source LLC
Chambersburg PA
CBHW050729180526
45159CB00003B/1175